AF327380

Dimitri Plemenos and Georgios Miaoulis

Visual Complexity and Intelligent Computer Graphics Techniques Enhancements

Studies in Computational Intelligence, Volume 200

Editor-in-Chief

Prof. Janusz Kacprzyk
Systems Research Institute
Polish Academy of Sciences
ul. Newelska 6
01-447 Warsaw
Poland
E-mail: kacprzyk@ibspan.waw.pl

Further volumes of this series can be found on our homepage: springer.com

Vol. 177. Dikai Liu, Lingfeng Wang and Kay Chen Tan (Eds.)
Design and Control of Intelligent Robotic Systems, 2009
ISBN 978-3-540-89932-7

Vol. 178. Swagatam Das, Ajith Abraham and Amit Konar
Metaheuristic Clustering, 2009
ISBN 978-3-540-92172-1

Vol. 179. Mircea Gh. Negoita and Sorin Hintea
Bio-Inspired Technologies for the Hardware of Adaptive Systems, 2009
ISBN 978-3-540-76994-1

Vol. 180. Wojciech Mitkowski and Janusz Kacprzyk (Eds.)
Modelling Dynamics in Processes and Systems, 2009
ISBN 978-3-540-92202-5

Vol. 181. Georgios Miaoulis and Dimitri Plemenos (Eds.)
Intelligent Scene Modelling Information Systems, 2009
ISBN 978-3-540-92901-7

Vol. 182. Andrzej Bargiela and Witold Pedrycz (Eds.)
Human-Centric Information Processing Through Granular Modelling, 2009
ISBN 978-3-540-92915-4

Vol. 183. Marco A.C. Pacheco and Marley M.B.R. Vellasco (Eds.)
Intelligent Systems in Oil Field Development under Uncertainty, 2009
ISBN 978-3-540-92999-4

Vol. 184. Ljupco Kocarev, Zbigniew Galias and Shiguo Lian (Eds.)
Intelligent Computing Based on Chaos, 2009
ISBN 978-3-540-95971-7

Vol. 185. Anthony Brabazon and Michael O'Neill (Eds.)
Natural Computing in Computational Finance, 2009
ISBN 978-3-540-95973-1

Vol. 186. Chi-Keong Goh and Kay Chen Tan
Evolutionary Multi-objective Optimization in Uncertain Environments, 2009
ISBN 978-3-540-95975-5

Vol. 187. Mitsuo Gen, David Green, Osamu Katai, Bob McKay, Akira Namatame, Ruhul A. Sarker and Byoung-Tak Zhang (Eds.)
Intelligent and Evolutionary Systems, 2009
ISBN 978-3-540-95977-9

Vol. 188. Agustín Gutiérrez and Santiago Marco (Eds.)
Biologically Inspired Signal Processing for Chemical Sensing, 2009
ISBN 978-3-642-00175-8

Vol. 189. Sally McClean, Peter Millard, Elia El-Darzi and Chris Nugent (Eds.)
Intelligent Patient Management, 2009
ISBN 978-3-642-00178-9

Vol. 190. K.R. Venugopal, K.G. Srinivasa and L.M. Patnaik
Soft Computing for Data Mining Applications, 2009
ISBN 978-3-642-00192-5

Vol. 191. Zong Woo Geem (Ed.)
Music-Inspired Harmony Search Algorithm, 2009
ISBN 978-3-642-00184-0

Vol. 192. Agus Budiyono, Bambang Riyanto and Endra Joelianto (Eds.)
Intelligent Unmanned Systems: Theory and Applications, 2009
ISBN 978-3-642-00263-2

Vol. 193. Raymond Chiong (Ed.)
Nature-Inspired Algorithms for Optimisation, 2009
ISBN 978-3-642-00266-3

Vol. 194. Ian Dempsey, Michael O'Neill and Anthony Brabazon (Eds.)
Foundations in Grammatical Evolution for Dynamic Environments, 2009
ISBN 978-3-642-00313-4

Vol. 195. Vivek Bannore and Leszek Swierkowski
Iterative-Interpolation Super-Resolution Image Reconstruction:
A Computationally Efficient Technique, 2009
ISBN 978-3-642-00384-4

Vol. 196. Valentina Emilia Balas, János Fodor and Annamária R. Várkonyi-Kóczy (Eds.)
Soft Computing Based Modeling in Intelligent Systems, 2009
ISBN 978-3-642-00447-6

Vol. 197. Mauro Birattari
Tuning Metaheuristics, 2009
ISBN 978-3-642-00482-7

Vol. 198. Efrén Mezura-Montes (Ed.)
Constraint-Handling in Evolutionary Optimization, 2009
ISBN 978-3-642-00618-0

Vol. 199. Kazumi Nakamatsu, Gloria Phillips-Wren, Lakhmi C. Jain, and Robert J. Howlett (Eds.)
New Advances in Intelligent Decision Technologies, 2009
ISBN 978-3-642-00908-2

Vol. 200. Dimitri Plemenos and Georgios Miaoulis
Visual Complexity and Intelligent Computer Graphics Techniques Enhancements, 2009
ISBN 978-3-642-01258-7

Dimitri Plemenos and Georgios Miaoulis

Visual Complexity and Intelligent Computer Graphics Techniques Enhancements

 Springer

Professor Dimitri Plemenos
Universite de Limoges
Laboratoire XLIM
83, rue d'Isle
87000 Limoges
France
E-mail: plemenos@unilim.fr

Professor Georgios Miaoulis
Technological Education Institute of Athens
Department of Computer Science
Ag. Spyridonos
Egaleo, 122 10 Athens
Greece

ISBN 978-3-642-01258-7   e-ISBN 978-3-642-01259-4

DOI 10.1007/978-3-642-01259-4

Studies in Computational Intelligence   ISSN 1860949X

Library of Congress Control Number: Applied for

Typeset & *Cover Design:* Scientific Publishing Services Pvt. Ltd., Chennai, India.

Printed in acid-free paper

9 8 7 6 5 4 3 2 1

springer.com

Preface

Even if research in computer graphics is well developed and spectacular progresses have been accomplished during these last years, several drawbacks remain, especially in some areas, neglected for a long time by computer graphics researchers. An example of such an area is research and development of tools permitting to make easier the work of the user of high quality computer graphics results.

So, high quality scene modeling techniques have been developed but they are difficult to use by a scene designer, because their use requires low level details and does not correspond to an intuitive designing walk. Intelligent scene design is a way to simplify scene modeling. Another example is the lack of intelligent tools for assisting scene rendering. It is obvious that, even if the rendering of a scene is very realistic, it is useless if it is impossible to understand the scene from its image, due to a bad choice of camera position. The user would appreciate to have an intelligent tool allowing automatic estimation of a good camera position.

In this book, three main notions will be used in our search of improvements in various areas of computer graphics: Artificial Intelligence, Viewpoint Complexity and Human Intelligence. Several Artificial Intelligence techniques are used in presented intelligent scene modelers, mainly declarative ones. Among them, the mostly used techniques are Expert systems, Constraint Satisfaction Problem resolution and Machine-learning. The notion of viewpoint complexity, that is complexity of a scene seen from a given viewpoint, will be used in improvement proposals for a lot of computer graphics problems like scene understanding, virtual world exploration, image-based modeling and rendering, ray tracing and radiosity. Very often, viewpoint complexity is used in conjunction with Artificial Intelligence techniques like Heuristic search and Problem resolution. The notions of artificial Intelligence and Viewpoint Complexity may help to automatically resolve a big number of computer graphics problems. However, there are special situations where is required to find a particular solution for each situation. In such a case, human intelligence has to replace, or to be combined with, artificial intelligence. Such cases, and proposed solutions will also be presented in this book.

The first chapter of the book briefly presents the main tools which will be used in techniques described in the remaining chapters, that is Artificial Intelligence and Viewpoint Complexity. Chapter 2 is dedicated to intelligent scene modeling (mainly declarative modeling) and to artificial intelligence techniques used to make possible such a scene modeling. Chapter 3 describes viewpoint complexity-based methods, using artificial intelligence techniques to automatically compute a good point of view, allowing understanding of a 3D scene. In chapter 4, virtual world exploration techniques are described. They are also based on viewpoint complexity and artificial intelligence and permit to understand complex scenes, which cannot be understood from only one image. Possible integration of light in scene understanding techniques is described in chapter 5, together with techniques, based on viewpoint complexity and improving traditional methods in various areas of computer graphics or its applications, like image-based modeling and rendering, ray-tracing or molecular rendering. Chapter 6 is dedicated to viewpoint complexity-based methods, permitting to improve traditional Monte Carlo radiosity computation techniques. In chapter 7 some cases, where artificial intelligence or viewpoint complexity-based techniques cannot automatically give satisfactory solutions, are presented. Data visualization-like techniques, based in human intelligence, are proposed for such cases.

The major part of the techniques presented in this book were studied and implemented under our direction. However, other techniques, realized in other research laboratories, are presented too. Most of them were developed in the University of Girona (Spain), and directed by Mateu Sbert.

Dimitri Plemenos
Georgios Miaoulis

Contents

1 Tools for Intelligent Computer Graphics

Abstract. In this chapter two important tools, allowing intelligent computer graphics, will be briefly described. The first one is Artificial Intelligence, while the second tool is Visual Complexity. A review of Artificial Intelligence techniques, which may be used to improve traditional methods in specific areas of Computer Graphics, is first presented in this chapter. After a definition of the main principles of Artificial Intelligence, three main areas of Artificial Intelligence are briefly described. *Problem resolution* gives general methods, which could be applied to resolve problems, especially heuristic search. Two specific sub-areas of problem resolution techniques are detailed: strategy computer games and the Constraint Satisfaction Problem. Expert Systems are able to advantageously replace human experts in various application areas. The main principles of Expert Systems are briefly presented. Finally, machine-learning techniques, mainly based on Neural Networks and Genetic Algorithms, are described.

Keywords. Artificial Intelligence, Computer Graphics, Problem Resolution, Strategy Games, Constraint Satisfaction Problem, Expert Systems, Machine-learning, Neural Networks, Genetic Algorithms.

1.1 Introduction

Computer graphics is one of the three areas of graphics processing techniques, the two others being Image processing and Pattern recognition. In a general manner, computer graphics includes (fixed or animated) scene modeling and (fixed or animated) scene visualization, this visualization being photo-realistic or not.

The two main phases of computer graphics, scene modeling and visualization, have been first developed independently of each other. Scene visualization techniques have been developed before scene modeling techniques in the years 1970, because the processing power of computers was then insufficient to process very complex scenes. So, it was not really a problem to create the simple scenes required by the rendering algorithms and adapted to the processing power of computers. During this period, very interesting visualization algorithms have been developed.

When, at the end of years 70, computers became more powerful, people discovered that it was then possible to process complex scenes with the existing

D. Plemenos and G. Miaoulis: Visual Complexity, SCI 200, pp. 1–25.

computers and algorithms but it was not possible to get complex scenes. Computer graphics researchers discovered that it is difficult to empirically design a scene by giving only coordinates of points or equations of curves or surfaces. Research on scene modeling has then begun and several models have been proposed in order to improve scene modeling. Currently, there exist well defined geometric models, used into powerful scene modelers, and the design of a scene is easier than ten or twenty years ago.

Even if today's scene modeling and rendering techniques are very powerful, there exist cases where the available tools are not entirely satisfactory, as well in scene modeling as in scene visualization and rendering. In these cases, the use of Artificial Intelligence techniques can improve the modeling and rendering processes.

The two main areas of computer graphics, scene modeling and scene visualization are currently well developed and allow to create and display rather complex scenes with a high degree of realism. However, several problems remain and artificial intelligence techniques could give a satisfactory solution to these problems. We are going here to give some examples of non resolved problems in computer graphics.

Scene modeling

Traditional geometric modeling is not well adapted to CAD. The main reasons of this are [5]:

- the lack of abstraction levels in descriptions, which renders some information difficult to obtain,
- the impossibility to use approximate or imprecise descriptions to express imprecise mental images of the user. The user of a scene modelling system would like to express high level properties of a desired scene and would like to let the modeller construct all of the scenes verifying these properties.

Scene understanding

Complex scenes are difficult to understand, especially scenes found on the web, because the scenes are three-dimensional and the screen two-dimensional and it is difficult to reach manually a good view position allowing to well understand a scene.

Radiosity

In Monte Carlo techniques, used to compute radiosity, an important number of randomly chosen rays are shot from each patch of the scene. These rays permit to regularly sample the scene and to diffuse the energy of each patch. Unfortunately, regular sampling is not always well adapted, because most of the scenes are not uniformly complex. Some regions of a scene can be complex enough while others can be very simple (e.g. a wall). How estimate the complexity of a region of a scene and how take it into account in order to improve classical Monte Carlo techniques ?

Other problems

Other important problems could obtain a satisfactory solution with artificial intelligence techniques: placement of a light source according to the expected results; design of scenes with specific properties, to be used to test new rendering methods, etc.

So, Artificial Intelligence techniques will be at the center of this book, where will try to explain how they may be used to improve traditional techniques of Computer Graphics. For this reason, this chapter will be partially dedicated to a short presentation of main Artificial Intelligence areas useful in Computer Graphics.

1.2   Elements of Artificial Intelligence for Computer Graphics

Artificial Intelligence is a vast computer science area proposing various techniques for writing "intelligent" programs, that is programs able to think and to reason like a human being [1, 2, 3].

The main characteristics of Artificial Intelligence programs are:

- **Manipulation of concepts**. Artificial Intelligence programs are able to process concepts, not only numerical data.

- **Use of heuristics**. Artificial Intelligence programs often use heuristic methods for resolving problems where no solution may be obtained by known algorithms.

- **Knowledge representation**. Unlike in other programs, the knowledge is often explicitly represented in Artificial Intelligence programs.

- **Acceptation of imprecise data**. Sometimes, the user of Artificial Intelligence has not complete data of his (her) problem. Artificial Intelligence programs must accept incomplete or imprecise data.

- **Acceptation of many solutions**. When data are imprecise, the resolution of a problem may give many solutions. As Artificial Intelligence programs accept imprecise data, they must accept many solutions too.

- **Acceptation of conflicting data**. Apparently conflicting data must be accepted by Artificial Intelligence programs. If data are really conflicting, no solution will be given to the problem.

- **Capability to learn**. Intelligent systems must integrate mechanisms of machine learning, in order to have a reasoning as close as possible to the human one.

In order to carry out these goals, Artificial Intelligence programs mainly use the following techniques:

- Information retrieval.

- Manipulation of knowledge..

- Abstraction.

In the rest of this chapter we are going to briefly present the most interesting, for us, areas of Artificial Intelligence, that is these areas which may help improve various methods used in Computer Graphics and find solutions for difficult to resolve problems. These areas are:

- Problem resolution, including strategy games and Constraint Satisfaction.

- Expert systems.

- Machine learning.

1.3 Problem resolution

1.3.1 General Considerations

There exist 4 general methods used in Artificial Intelligence programs in order to resolve problems, according to the kind of available information.

1. If there exists a formula giving the solution, apply this formula to find the solution.
2. Enumerate all possibilities for a solution and evaluate each possibility to see if it is a solution.
3. If it is impossible to enumerate all possibilities for a solution, make a heuristic search in order to find partial interesting solutions.
4. If the problem is too complex to be resolved with one of the previous methods, decompose it in a set of sub-problems and apply to each of them one of the four problem resolution methods (problem reduction).

Heuristic search does not guarantees that the best solution will be found but it offers a high probability to find a good solution. In a general manner, heuristic search applies the following steps , starting from an initial state.

1. Choose an action among the possible actions.
2. Apply the chosen action which modify the current state.
3. Evaluate the current state.
4. If the current state is a final state (close to a solution), stop the process. Otherwise, choose a new state and apply again the heuristic search process.

The choice of a new state can be made:
- By keeping the last obtained state.
- By backtracking and choice of the most promising action.

1.3.2 Strategy games

Problem resolution techniques give good results in strategy games like chess, draughts, etc.

The used technique is often the following:

a. Generate all potential moves for the next ply.
b. Evaluate each move.
c. Choose the best one.

The programs which play strategy games, use an *evaluation function* working on numerical values. Evaluation function is different for different games. Even for the same game, it is possible to imagine a lot of possible evaluation functions. Evaluation function takes generally into account various criteria, such as:

• Values of available pieces.
• Positions of pieces.
• etc.

Programs playing games for two players apply mainly two kinds of strategies in evaluation of possible next plies for a player.

The first strategy, named minmax (or minimax), tries to evaluate a possible ply of the player, taken into account the possible plies of the opponent player, up to a given level of depth. In this strategy it is supposed that each player will choose the best ply, according to the used evaluation function.

The purpose of the second strategy, named alpha-beta, is to reduce the evaluation cost by pruning sub-trees of the minmax research tree of possible plies, when it is possible to know that the player (or his opponent) will never choose his next ply on this tree.

1.3.3 Constraint satisfaction

A Constraint Satisfaction Problem (CSP) is made of:

1. A set of variables.
2. A set of constraints on the values of variables.
3. A set of possible values of variables, called the domain of variables. Variables may receive integer values.

Example:
Set of variables: x, y, z.
Constraints: x=y+z, y<z.
Domains: x in [1..4], y in [1..5], z in [1..4]

Resolution of a constraint satisfaction problem consists to find, for each variable, a value verifying all the constraints of the CSP.

Example:
For the following CSP:
Variables: x, y, z.
Constraints: x=y+z, y<z.
Domains: x in [1..4], y in [1..5], z in [1..4]

A solution is: x=3, y=1, z=2.

General resolution methods for constraint satisfaction problems contain 2 steps.

1. The enumeration step, where a new value is assigned to the current variable.
2. The propagation step, where the already known values of variables are propagated to all the constraints of the CSP.

These steps may be repeated in order to get a solution. When the current values of variables cannot give any solution, a backtracking is performed and the next value is given to the previous variable.

More precisely, the resolution process for a CSP is described below.

1.3.3.1 The Resolution Process

The resolution process for a CSP is an iterative process, each step being composed of two phases: the enumeration phase and the propagation phase.

1. The *enumeration phase* is made of two parts:
    *   Selection of a variable;

- Selection of a value for this variable.

It is possible to use various orders in selecting variables and values. Some selection orders give best results than others. The variable selection order giving the best results it the order which follows the hierarchical decomposition tree given in the description phase.

2. The *propagation phase* takes into account the value of the variable chosen in the enumeration phase and tries to reduce the constraint satisfaction problem by propagating this value. Propagation of the value of the selected variable can be made in different manners:
   - Propagation through the affected constraints of the already instantiated variables. In this method, called "look-back", the resolution engine checks if the new value is compatible with the previously chosen values of variables.
   - Propagation through all variables, even the not yet instantiated. This kind of propagation is called "look-ahead" and can avoid future conflicts.

   If the "look-ahead" technique performs arc-consistency tests only between an instantiated and a not yet instantiated variable, it is called "forward-checking". If arc-consistency tests can be performed between not yet instantiated variables that have no direct connection with instantiated variables, in this case, the method is called "partial look-ahead".

1.3.3.2 Constraint Logic Programming on Finite Domains – CLP(FD)

The CLP(FD) method was initially proposed by Pascal Van Hentenryck [15, 16]. Several other authors contributed to its development.

The CLP(FD) method is based on the use of only primitive constraints of the form X **in** r, where X is a variable taking its values in a finite domain and r is a range. A range is a dynamic expression of the current domain of a variable and uses references to other variables via four *indexicals*, **min** (the min of the domain of the variable), **max** (the max of the domain of the variable), **dom** (the domain of the variable) and **val** (the value of the variable). Arithmetic expressions using indexicals, integer values and arithmetic operators can be used to describe a range.

In the final set of primitive constraints to be processed, each variable appears only once under the form: X **in** r1&r2&...&rn, where r1, r2, ..., rn are ranges and & denotes the union of ranges. This final form may be automatically obtained by applying rewriting rules. For instance, constraint "X=Y+Z" will give the set of primitive constraints:

X **in** min(Y)+min(Z) .. max(Y)+max(Z)

Y **in** min(X)-max(Z) .. max(X)-min(Z)

Z **in** min(X)-max(Y) .. max(X)-min(Y)

During the resolution process, a value is given to a variable X in order to reduce its range to a single value. The solver is then looking for indexicals using X in the domains of all the other variables and updates them. The ranges of the variables using these indexicals is reduced and this reduction is propagated to other variables. If the range of a variable becomes empty, there are no solutions for the chosen value of X and another value has to be tried.

With primitive constraints, various propagation strategies can be applied, depending on indexicals taken into account. Thus:

- Full look-ahead is obtained by using the indexical **dom**,
- Partial look-ahead is obtained by using only indexicals **min** and **max**. In this way, holes are not transmitted.
- Forward checking by using the indexical **val**, delaying the check of the constraint until the involved variable is ground.

1.4 Expert Systems

An expert system is made to replace an expert in a given area. Usually an expert system is composed of:

- A set of **facts** (knowledge always true),

- A set of **rules** (reasoning on knowledge)

- An **inference engine** able to activate the reasoning mechanism by applying the rules to facts.

Fig. 1.1 illustrates the usual architecture of an expert system.

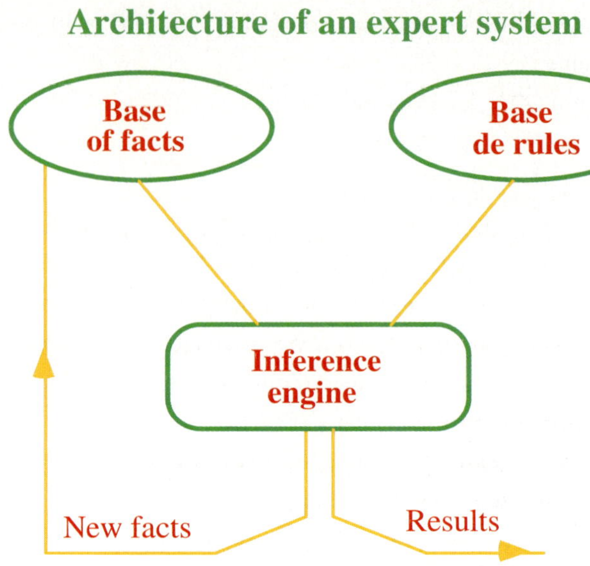

Architecture of an expert system

Fig. 1.1: Architecture of an expert system

1.4.1 Rules and inference engine of expert systems

The rules of an expert system are often of the form:

X1, X2, …, Xn --->Y
where:
Xi are the conditions of the rule;
Yis the conclusion.

The rules of an expert system may contain variables or not.

An expert system with rules containing variables is more powerful than expert systems without variables.

The inference engine of an expert system applies rules in order to answer questions like:

- Is proposal P true or false?
- For which values of variables is proposal P true?

1.4.2 Main strategies for applying rules

An inference engine may use one of two main strategies in the manner to apply the rules.

- **The forward reasoning strategy.**

When the forward reasoning strategy is applied, a rule X1, X2, …, Xn --->Y of the base of rules is interpreted as follows:

If X1 **and** X2 **and** … **and** Xn are true, Y is true too.

- **The backward reasoning strategy.**

When the backward reasoning strategy is applied, a rule X1, X2, …, Xn --->Y of the base of rules is interpreted as follows:

If we wish Y to be true, it is enough to verify that X1 **and** X2 **and** … **and** Xn are true.

In the forward reasoning strategy, the inference engine starts from the facts and applies only rules whose all conditions are facts. The process stops when a reply to the initial question is found or when the base of facts is saturated.

In the backward reasoning strategy, the inference engine starts from the question to process and applies only rules whose conclusions are this question. If all the conditions of one of these rules is verified, the question is verified too.

1.5 Machine-learning

Learning is one of the main components of human reasoning. A human is able to learn and to integrate new experiences to his reasoning. Is it possible to imagine and to design systems able to learn? Humans have always wanted to construct machines able to learn, because learning is the property that characterizes better the human intelligence.

One of the most useful kinds of machine learning is learning from examples, where examples given by the teacher help the system to make generalizations and to deduce the good properties of objects.

After learning, the system can work as a filter accepting only objects with good properties.

In this section we are going to present two machine-learning paradigms, *neural networks* and *genetic algorithms*.

1.5.1 Neural networks

A neural network is a set of artificial neurons connected by links. A weight is associated to each link. In Fig. 1.2 one can see an example of neural network.

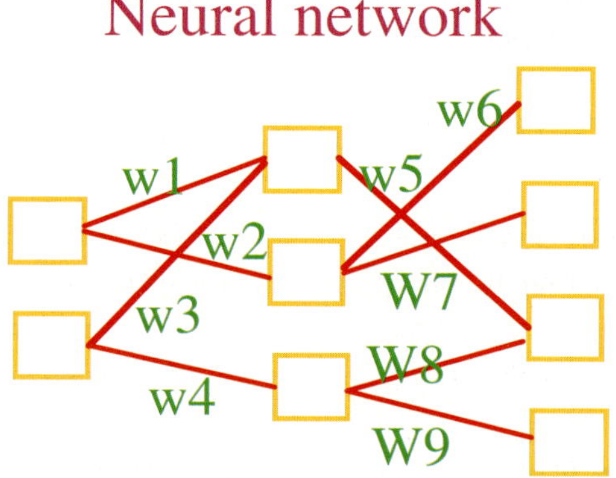

Fig.1.2: Example of neural network

Each artificial neuron can transmit an information to its neighbors, only if the weights of the corresponding links are greater than a threshold value.

An artificial neuron is characterized by several inputs, one output and one activation function (see Fig. 1.3). To each input of the neuron is associated a weight.

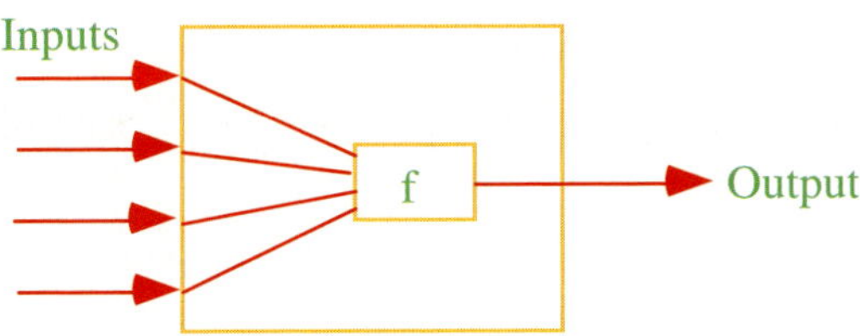

Fig. 1.3: Structure of an artificial neuron.

An activation f(S) function is associated to each artificial neuron, where S is the weighted sum of all inputs of the neuron.

$$\text{If } S = \sum_{i=1}^{n} e_i w_i$$

The activation function of the neuron will be f(S).
Here are some examples of usual activation functions:

1.   f : identity function

$$f(S) = S = \sum_{i=1}^{n} e_i w_i$$

2.   f : function with threshold value

$$f(S) = 0 \text{ if } S \leq s$$
$$f(S) = 1 \text{ if } S > s$$

3.   f : sigmoid function $f(S) = \dfrac{1}{1 + e^{-k(S-\sigma)}}$

with k constant and σ: threshold.

A learning rule is associated to each neural network. This rule defines the manner in which the weights of the network are modified, depending on examples presented to it.

The machine learning phase of a neural network may be described as follows:

Action MachineLearning

   for all examples $(e_1, e_2, \ldots, e_n)$ **do**

$$S(t) \leftarrow \sum_{i=1}^{n} e_i w_i$$

   for all wi **do**

$$w_i(t) = w_i(t-1) + v \times g(e_i(t), e_i(t-1), S(t), S(t-1))$$

   end for

   end for

End Action

Where t is the rank of the current example and $(e_1, e_2, \ldots, e_n)$ are the inputs for the current example.

1.5.2 Genetic algorithms

Genetic algorithms are optimization methods based on evolution theories. Machine-learning with genetic algorithms occurs at the end of a selection process applied to a big number of individuals (or chromosomes). Each individual is characterized by an importance degree. From the initial generation, new generations are created by application of genetic operations. Usually, individuals are represented by bit strings. The usual genetic operations are the following.

- **Selection**. This operation allows to select individuals to which the other genetic operations will be applied. Selection takes into account the importance degree of individuals. The probability to select an individual is function of its importance.

- **Copy**. This operation is applied to an individual an produces another individual. It consists to copy the selected individual to the next generation.

- **Crossover**. This operation is applied to two selected individuals of the current generation and produces two individuals for the next generation. A binary position is randomly chosen and each selected individual is divided in two parts, according to the chosen binary position. The first part of the first individual is completed with the second part of the second one and the first part of the second individual is completed with the first part of the first one, in order to create two new individuals.

Example:
Chosen individuals: 11010010 and 10001100
Randomly chosen binary position: 2
Two sub-strings are obtained from each individual:
Individual 1: 11 and 010010
Individual 2: 10 and 001100
Two new individuals are obtained by crossover for the next generation:
Individual 1: 11001100
Individual 2: 10010010

- **Mutation**. This operation is applied to a selected individual of the current generation and produces another individual for the next generation. A binary position is randomly chosen and, if the corresponding bit of the individual is 1, it is changed to 0 and vice versa.

Example:
Chosen individual: 11010010
Randomly chosen binary position: 3
New individual for the next generation: 11110010

A different application ratio is associated to each genetic operation. Usually, the application ratio is 0.1 for crossover and 0.001 for mutation. The stop criterion for a genetic algorithm may be a previously fixed number of steps or obtainment of individuals with an importance degree greater than a threshold value.

The solution of a Genetic algorithms based machine-learning is represented by the best individual (that is the individual with the best importance degree) or the best group of individuals, of the final population.

1.6 Artificial Intelligence and Computer Graphics

In this chapter we have briefly seen the main Artificial Intelligence techniques which may be used to improve Computer Graphics methods. Which areas of Computer Graphics could be improved by introduction, of Artificial Intelligence techniques? We think that these techniques could improve many areas of Computer Graphics. For the moment, we will briefly describe Computer Graphics areas where Artificial Intelligence techniques are already used.

Declarative scene modelling.
The purpose of this area of Computer Graphics is to emancipate the designer from some drawbacks of traditional geometric scene modelling by allowing him (her) to use imprecise or incomplete descriptions in designing scenes. The main Artificial Intelligence techniques used in declarative modelling are: expert systems and constraint satisfaction techniques, in order to build an intelligence scene generation engine; Machine-learning techniques (neural networks, genetic algorithms) in order to improve performances of the scene generation engine.

Visual scene understanding an virtual world exploration.
The problem is to permit a user to understand a scene from its image on the screen of a computer, or by exploring the scene with a virtual camera. The main Artificial Intelligence techniques used for this task are: Heuristic search for the search of good camera positions; Strategy games techniques in choosing the next camera position during scene exploration.

Behavioral animation.
In virtual reality applications, autonomous virtual agents are entities which own their proper behavior and to which individual tasks are assigned. In order to help these agents to take decisions, various Artificial Intelligence techniques are used, such as expert systems, neural networks or genetic algorithms.

Getting optimal triangulated scenes.
Triangulated scenes are very often used in Computer Graphics. Triangulation of scenes is a too heavy task to be done manually. So, this task is usually performed automatically by programs. The problem is that this automatic triangulation is not always optimal. Intelligent scene triangulation programs use mainly genetic algorithms which help to get optimal triangulated scenes.

Image-based modelling and rendering.
Image-based modelling and rendering is a solution for reducing the cost of scene rendering. Rendering of a scene is replaced by images, already computed and stored or obtained by interpolation from existing ones. The problem is to optimize the camera positions used to compute the stored images. Artificial Intelligence techniques like heuristic search or strategy games techniques may be used to automatically compute optimal camera positions.

Inverse lighting.
The inverse lighting problem is a difficult one because one has to ask the question "Where do I put light source in order to get a given lighting". Declarative modelling, combined with fuzzy logic, permit to give satisfactory solutions for this kind of problem.

In the following chapters a detailed presentation of some intelligent techniques, which are already used or could be used in Computer Graphics, will be made.

1.7 What is viewpoint complexity in Computer Graphics?

It is generally admitted that there are scenes which are considered more complex than others. The notion of complexity of a scene is an intuitive one and, very often, given two different scenes, people are able to say which scene is more complex than the other one. Another problem is that it is not always clear what kind of complexity people is speaking about. Is it computational complexity, taking into account the computational cost of rendering; geometric complexity, taking into account the complexity of each element of the scene; quantitative complexity, depending on the number of elements of the scene?

We can informally define the *intrinsic complexity* of a scene as a quantity which:

1. Does not depend on the point of view;
2. Depends on:
  • The number of details of the scene.
  • The nature of details (convex or concave surfaces).

Some steps towards a formal definition of scene complexity had been presented in [13, 14]. Unlike intrinsic complexity, the *viewpoint (or visual) complexity* of a scene depends on the point of view. An *intrinsically complex* scene, seen from a particular point of view, is not necessarily visually complex. A first measure of the notion of viewpoint complexity of a scene from a point of view could be the number of visible details or, more precisely, the number of surfaces of the scene visible from this point of view. However, this definition of viewpoint complexity is not very satisfactory because the size of visible details is also important. Informally, we will define the viewpoint complexity of a scene from a given point of view as a quantity which depends on:

- The number of surfaces visible from the point of view.
- The area of visible part of each surface of the scene from the point of view.
- The orientation of each (partially) visible surface according to the point of view.
- The distance of each (partially) visible surface from the point of view.

An intuitive idea of viewpoint complexity of a scene is given in Fig. 1.4, where the visual complexity of scene (a) is less than the visual complexity of scene (b) and the visual complexity of scene (b) is less than the visual complexity of scene (c), even if each scene contains other non visible surfaces.

Given a point of view, a scene may be divided in several more or less visually complex regions from this point. The visual complexity of a (part or region of) scene from a point of view, as defined above, is mainly geometry-based as the elements taken into account are geometric elements. It would be possible to take into account other aspects of the scene, such as lighting, because lighting may modify the perception of a scene by a human user. However, only geometrical viewpoint complexity will be discussed in this chapter, but a study on influence of lighting in scene perception will be made in chapter 5.

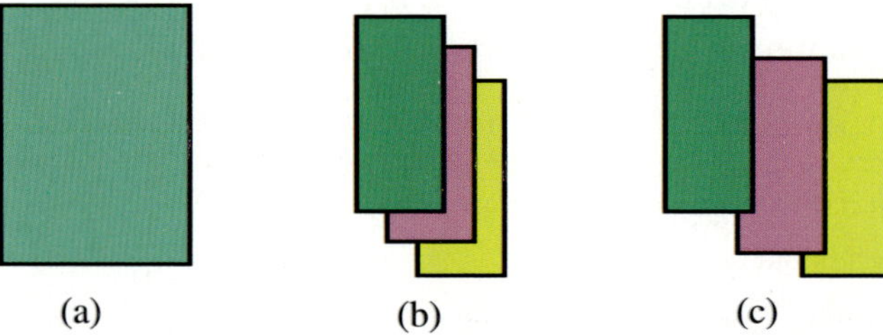

(a)   (b)   (c)

Fig. 1.4: Visual complexity of scene (c) is greater than visual complexity of scene (b); visual complexity of scene (b) is greater than visual complexity of scene (a).

1.8   How to compute visual complexity?

Following our definition of viewpoint complexity in this chapter, its calculation depends on the number of visible surfaces, the area of the visible part of each surface and the distance and orientation of each (partially) visible surface, according to the point of view. A linear combination of these two quantities would give an accurate enough measure of visual complexity. The most important problem is the way to compute the number of visible surfaces and the visible projected area of each surface. The used method may depend on some constraints of the application using this information. Some applications require real time calculation whereas for others the calculation time is not an important constraint. For some applications, it is very important to have accurate visual complexity estimation and for others a fast approximate estimation is sufficient.

It is easy to see that the visible part, orientation and distance of (partially) visible surfaces from the point of view can be accurately approximated by the projection of the visible parts of the scene on the unitary sphere centred on the point of view. This approximation will be used in this section to estimate the visual complexity of a scene from a point of view.

1.8.1 Accurate visual complexity estimation

The most accurate estimation of the visual complexity of a scene can be obtained by using a hidden surface removal algorithm, working in the user space and explicitly computing the visible part of each surface of the scene. Unfortunately, it is rarely possible in practice to use such an algorithm because of the computational complexity of this kind of algorithms. For this reason, less accurate but also less complex methods have to be used.

A method proposed in [6, 7, 12] and used in [9] permits to use hardware accelerated techniques in order to decrease the time complexity of estimation. This method uses image analysis to reduce the computation cost. Based on the use of the OpenGL graphical library and its integrated z-buffer, the technique used is the following.

If a distinct colour is given to each surface of the scene, displaying the scene using OpenGL allows to obtain a histogram (Fig. 1.5) which gives information on the number of displayed colours and the ratio of the image space occupied by each colour.

As each surface has a distinct colour, the number of displayed colours is the number of visible surfaces of the scene from the current position of the camera. The ratio of the image space occupied by a colour is the area of the projection of the visual part of the corresponding surface. The sum of these ratios is the projected area of the visible part of the scene. With this technique, the two visual complexity criteria are computed directly by means of an integrated fast display method.

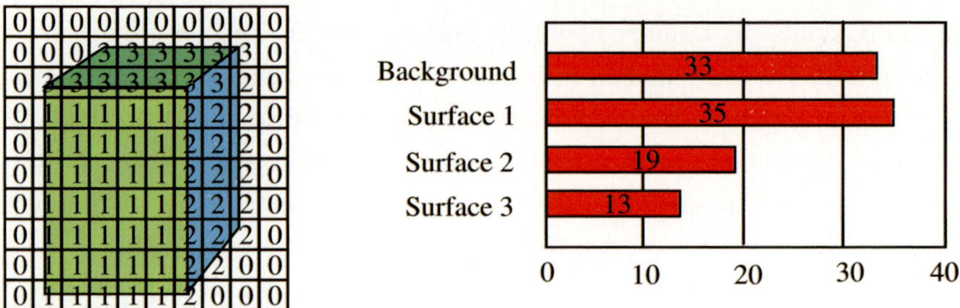

Fig. 1.5: Fast computation of number of visible surfaces and area of projected visual part of the scene by image analysis.

The visual complexity of a scene from a given viewpoint can now be computed by a formula like the following one:

$$C(V) = \frac{\sum\limits_{i=1}^{n} \left[\dfrac{P_i(V)}{P_i(V)+1}\right]}{n} + \frac{\sum\limits_{i=1}^{n} P_i(V)}{r}$$

where: C(V) is the visual complexity of the scene from the view point V,

Pi(V) is the number of pixels corresponding to the polygon number i in the image obtained from the view point V,

r is the total number of pixels of the image (resolution of the image),

n is the total number of polygons of the scene.

In this formula, [a] denotes the smallest integer, greater than or equal to a.

Another method to compute viewpoint complexity has been proposed in [8, 10, 11], based on information theory. A new measure is used to evaluate the amount of information captured from a given point of view. This measure is called *viewpoint entropy*. To define it, the authors use the relative area of the projected faces over the sphere of directions centred in the point of view (Fig. 1.6).

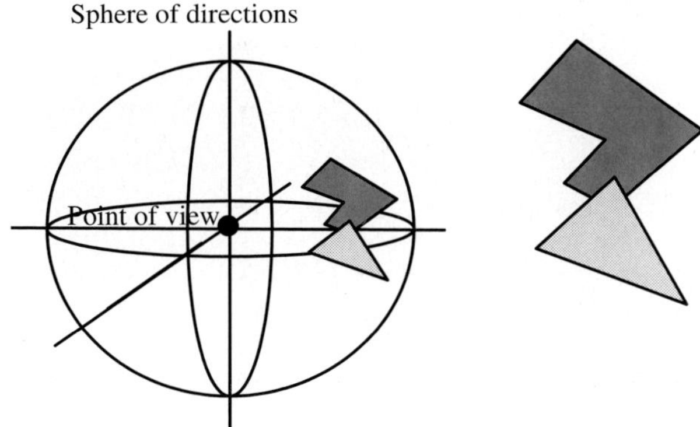

Fig. 1.6: the projected areas of polygons are used as probability distribution of
the entropy function

In this method, the viewpoint complexity of a scene from a given point of view
is approached by computing the viewpoint entropy. The viewpoint entropy is
given by the formula:

$$H(S,P) = - \sum_{i=0}^{N_f} \frac{A_i}{A_t} \log \frac{A_i}{A_t}$$

where P is the point of view, Nf is the number of faces of the scene S, Ai is the
projected area of the face i and At is the total area covered over a sphere centred
on the point of view.

The maximum entropy is obtained when a viewpoint can see all the faces with
the same relative projected area Ai/At. The best viewpoint is defined as the one
that has the maximum entropy.

Fig. 1.7: point of view based on viewpoint entropy (Image P. P. Vazquez)

To compute the viewpoint entropy, the authors use the technique explained above, based on the use of graphics hardware using OpenGL.

The selection of the best view of a scene is computed by measuring the viewpoint entropy of a set of points placed over a sphere that bounds the scene. The point of maximum viewpoint entropy is chosen as the best one. Fig. 1.7 presents an example of results obtained with this method.

Both methods have to compute the number of visible surfaces and the visible projected areas by using the technique described above.

Kamada and Kawai [4] use another, less accurate, kind of viewpoint complexity to display 3D scenes.

1.8.2 Fast approximate estimation of visual complexity

In some cases, accurate viewpoint complexity estimation is not requested, either because of need of real time estimation of the visual complexity or because a less accurate estimation is enough for the application using the visual complexity.

In such a case, it is possible to roughly estimate the viewpoint complexity of a scene from a given point of view, as follows:
A more or less great number of rays are randomly shot from the point of view to the scene and intersections with the surfaces of the scene are computed. Only intersections with the closest to the point of view surfaces are retained (Fig. 1.8).

Now, we can approximate the quantities used in viewpoint complexity calculation. We need first to define the notion of visible intersection. A *visible intersection* for a ray is the closest to the point of view intersection of the ray with the surfaces of the scene.

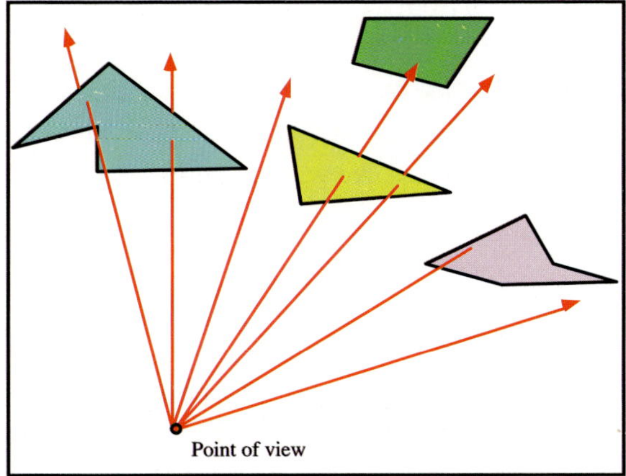

Fig. 1.8: Approximated estimation of viewpoint complexity

- *Number of visible surfaces* = number of surfaces containing at least one visible intersection with a ray shot from the point of view.
- *Visible projected area of a surface* = number of visible intersections on the surface.
- *Total visible projected area* = number of visible intersections on the surfaces of the scene.
- *Total projected area* = total number of rays shot.

The main interest of this method is that the user can choose the degree of accuracy, which depends on the number of rays shot.

1.9 Conclusion

In this first chapter we have presented two important tools, allowing improvements in various computer graphics areas. The first tool was Artificial Intelligence, where a lot of techniques (problem resolution, expert systems, neural networks, genetic algorithms, etc.) permit to write intelligent programs. Of course, this presentation is a partial one because we have presented only Artificial Intelligence techniques which will be used in next chapters.

The second tool was the notion of viewpoint complexity. An informal definition of this notion was given, as well as various methods to compute the viewpoint complexity of a scene, seen from a given viewpoint.

These tools will be used in the remaining chapters of this volume, where will be presented techniques permitting to give satisfactory solutions to many computing graphics problems, difficult to resolve with traditional methods.

References

1. Rich, E., Knight, K.: Artificial Intelligence. McGraw-Hill, New York (1991)
2. Plemenos, D.: Artificial Intelligence and Expert Systems. Course handbook. University of Limoges (1999)
3. Goldberg, D.E.: Genetic Algorithms. Addison-Wesley, Reading (1991)
4. Kamada, T., Kawai, S.: A Simple Method for Computing General Position in Displaying Three-dimensional Objects. Computer Vision, Graphics and Image Processing 41 (1988)
5. Plemenos, D.: A contribution to the study and development of scene modeling, generation and visualization techniques. The MultiFormes project. Professorial dissertation, Nantes, France (November 1991)
6. Plemenos, D., Benayada, M.: Intelligent display in scene modeling. New techniques to automatically compute good views. In: GraphiCon 1996, Saint Petersburg (July 1996)
7. Barral, P., Dorma, G., Plemenos, D.: Visual understanding of a scene by automatic movement of a camera. In: GraphiCon 1999, Moscow, Russia, August 26 - September 3 (1999)
8. Sbert, M., Feixas, M., Rigau, J., Castro, F., Vazquez, P.-P.: Applications of the information theory to computer graphics. In: International Conference 3IA 2002, Limoges, France, May 14-15 (2002)
9. Barral, P., Dorme, G., Plemenos, D.: Scene understanding techniques using a virtual camera, Short paper. In: Eurographics 2000, Interlagen, Switzerland, August 20-25 (2000)
10. Vazquez, P.-P., Sbert, M.: Automatic indoor scene exploration. In: Proceedings of the International Conference 3IA 2003, Limoges, France, May 14-15 (2003)
11. Vazquez, P.-P.: On the selection of good views and its application to computer graphics. PhD Thesis, Barcelona, Spain, May 26 (2003)
12. Dorme, G.: Study and implementation of 3D scene understanding techniques. PhD thesis, University of Limoges, France, French (June 2001)

13. Plemenos, D., Sbert, M., Feixas, M.: On viewpoint complexity of 3D scenes. In: STAR Report International Conference GraphiCon 2004, Moscow, Russia, September 6-10, 2004, pp. 24–31 (2004)
14. Sbert, M., Plemenos, D., Feixas, M., Gonzalez, F.: Viewpoint Quality: Measures and Applications. In: Computational Aesthetics in Graphics, Visualization and Imaging, Girona, pp. 1–8 (May 2005)
15. van Hentenryck, P.: Constraint satisfaction in logic programming. Logic Programming Series. MIT Press, Cambridge (1989)
16. Diaz, D.: A study of compiling techniques for logic languages for programming by constraints on finite domains: the clp(FD) system

2 Intelligent scene modeling

Abstract. Well known drawbacks of traditional geometric scene modelers make difficult, and even sometimes impossible, intuitive design of 3D scenes. In this chapter we will discuss on intelligent scene modeling, that is modeling using intelligent techniques during the designing process and thus allowing intuitive scene design. The main presented modeling technique is declarative modeling [1], with a special attention to the used Artificial Intelligence techniques, in order to produce efficient and easy to use scene modelers.

Keywords. Declarative scene modeling, Geometric scene modeling, Constraint Satisfaction Problem, Machine-learning, Neural networks, Genetic algorithms, Fuzzy Logic, Scene understanding.

2.1 Introduction

Nowadays is generally admitted that traditional 3D scene modelers are badly adapted to Computer Aided Design. Scene modeling is a very difficult task and the usual tools proposed by geometric scene modelers are not able to satisfy the needs of intuitive design. in computer graphics, as traditional geometric modelers are not well adapted to computer aided design. With most of the current modeling tools the user must have quite precise idea of the scene to design before using a modeler to achieve the modeling task. In this manner, the design is not really a computer aided one, because the main creative ideas have been elaborated without any help of the modeler.

The main problem with most of the current scene modelers is that the designing process they allow is not really a computer aided one, as they need, very soon during the designing process, low-level details which are not important in the creative phase of design. This is due to the lack of levels of abstraction allowing the user to validate general ideas before resolve low-level problems.

Another problem is that imprecise or incomplete descriptions are generally not accepted by traditional geometric scene modelers. However, such descriptions are very useful, especially at the beginning of the designing process, as the general mental idea of the designer is not yet precise enough and he (she) would like to test various solutions proposed by the modeler from an initial imprecise description.

D. Plemenos and G. Miaoulis: Visual Complexity, SCI 200, pp. 27–64.

Declarative scene modeling tries to give intuitive solutions to this kind of problem by using Artificial Intelligence techniques which allow the user to describe high level properties of a scene and the modeler to give all the solutions corresponding to imprecise properties. Some declarative modelers even allow using various levels of detail in one description, permitting to describe more precisely the most interesting parts of a scene and, temporarily, the other parts.

The remaining of this chapter will be mainly dedicated to declarative scene modeling. The main used Artificial Intelligence techniques used to implement declarative scene modeling will be presented and illustrated via the main designing steps of four declarative scene modelers.

2.2 Declarative scene modeling

Declarative modeling [1, 2, 3, 5, 38] in computer graphics is a very powerful technique allowing to describe the scene to be designed in an intuitive manner, by only giving some expected properties of the scene and letting the modeler find solutions, if any, verifying these properties.

As the user may describe a scene in an intuitive manner, using common expressions, the described properties are often imprecise. For example, the user can tell the modeler that "the scene A must be put on the left of scene B". There exist several possibilities to put a scene on the left of another one. Another kind of imprecision is due to the fact that the designer does not know the exact property his (her) scene has to satisfy and expects some proposals from the modeler. So, the user can indicate that "the house A must be near the house B" without giving any other precision. Due to this lack of precision, declarative modeling is generally a time consuming scene modeling technique.

It is generally admitted that the declarative modeling process is made of three phases:
- the *description* phase, where the designer describes the scene;
- the *scene generation* phase, where the modeler generates one or more scenes verifying the description; and
- the *scene understanding* phase, where the designer, or the modeler, tries to understand a generated scene in order to decide whether the proposed solution is a satisfactory one, or not.

In the following sections of this chapter we will present the main Artificial Intelligence techniques used in each of the three phases of declarative scene modeling. Four declarative scene modelers will be mainly used to illustrate this presentation: PolyFormes [6, 7], MultiFormes [2], DE2MONS [24] and VoluFormes [21]. The main principles of these modelers are explained bellow. The reader could also see [45] for details on declarative scene modelers. Some other kinds of declarative or declarative-like modellers may be found in [15, 16, 30, 32].

2.2.1 The PolyFormes Declarative Scene Modeler

PolyFormes [6, 7] is the first experimental declarative modeler. It was developed in Nantes (France) at the end of 80's. Its first version was written in 1987 in Prolog. Other versions have been developed later in Pascal. The goal of the Poly-Formes declarative modeler is to generate all regular and semi-regular polyhedrons, or a part of the whole, according to the user's request.

2.2.2 The MultiFormes Declarative Scene Modeler

MultiFormes is a general purpose declarative scene modeler, initially developed in Nantes at the end of 80's. After 1992, new versions of the modeler were developed in Limoges (France). MultiFormes is based on a promising designing and modeling technique, *declarative modeling by hierarchical decomposition* (DMHD) [2, 3, 4, 12, 13, 14, 17, 18, 31, 34, 35, 38, 40, 41, 42, 43, 44]. The DMHD technique can be resumed as follows:

- If the current scene can be described using a small number of predefined high level properties, describe it.
- Otherwise, describe what is possible and then decompose the scene in a number of sub-scenes. Apply the DMHD technique to each sub-scene.

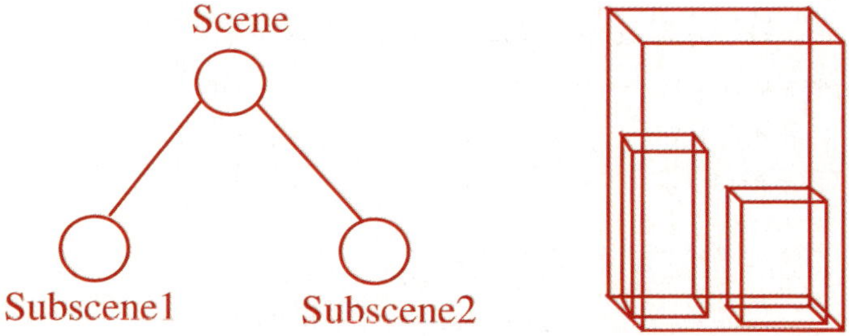

Fig. 2.1: The bounding boxes of the sub-scenes of a scene are inside the bounding box of the parent scene

The tree of the hierarchical description of a scene is used in the scene generation phase and allows scene generation in various levels of detail and reduction of
the generation's cost. To do this, the modeler uses a bounding box for each node
of the tree. This bounding box is the bounding box of the sub-scene represented by
the sub-tree whose the current node is the root. All bounding boxes of the children
nodes of a node are physically included in the bounding box of the parent node.
This property permits to detect very soon branches of the generation tree which
cannot be solutions. In Fig. 2.1, the spatial relation between the bounding boxes of
a scene and its sub-scenes is shown.

2.2.3 The DE2MONS Declarative Scene Modeler

The DE2MONS [24] declarative modeler is a general purpose modeler developed
in Toulouse (France), at the end of the years 1990. The main properties of
DE2MONS are:

- A multi modal interface,
- A generation engine limited to the placement of objects,
- A constraint solver able to process dynamic and hierarchical constraints.

The modeler uses a multi modal interface allowing descriptions by means of the
voice, the keyboard (natural language), a data glove or 3D captors informing the
system of the user's position. The description is translated in an internal model
made of linear constraints.

2.2.4 The VoluFormes Declarative Scene Modeler

VoluFormes [21] is a declarative modeler allowing the user to quickly define
boxes in the space whose purpose is to check the growth of forms. The modeller
was implemented in Nantes (France) at the middle of years 1990. It is made of two
modules:

- *Voluboites*, which allows to define boxes where the spatial control is performed.
- *Voluscenes*, which allows to use growth mechanisms applied to elementary
  germs and to create forms, taking into account the spatial control boxes.

Only Voluboites works in declarative manner.

2.3 The description phase in declarative scene modeling

In order to describe the mental idea the designer has of the scene to be designed, he (she) may generally use an interface allowing precise or imprecise descriptions. This description is then translated into an internal description model, which may be understood and processed by the modeler.

Two kinds of internal description models are mainly used, depending on the particular declarative modeler: the *set of rules and facts* model and *the set of arithmetic or geometric constraints* model.

2.3.1 Set of rules and facts

The external description given by the designer is translated into a set of rules and facts, which could be processed by the inference engine of an expert system.

This kind of external description model was used by PolyFormes[6, 7], the first implemented experimental declarative modeler, which was developed in Nantes (France). Its first version was written in 1987 in Prolog. Other versions have been developed later in Pascal. The goal of the PolyFormes declarative modeler is to generate all regular and semi-regular polyhedrons, or a part of the whole, according to the user's request.

Requests may be more or less precise and are expressed using dialog boxes. This initial description is then translated in an internal model which will be used during the generation process. This internal description model is a knowledge base, made of a knowledge base, containing rules and facts. Examples of rules contained in the PolyFormes knowledge base are the following:

R1: If F1 is a face and a is the dihedral angle between faces F1 and F2, then F2 is a neighboring face of F1.

R2: If the types of all faces attached to a vertex are known, then all dihedral angles between these faces are known.

In the first version of MultiFormes the internal description model was also a knowledge base made of a set of rules and a set of facts.

```
Workspace (0, 0, 0, 10, 10, 10) ->;

House(x) ->
    CreateFact(x)
```

```
        Habitation(x1)
        Garage(x2)
        PasteLeft(x1,x2);

    Habitation(x) ->
        CreateFact(x)
        HigherThanWide(x)
        HigherThanDeep(x)
        Roof(x1)
        Walls(x2)
        PutOn(x1,x2);

    Garage(x) ->
        CreateFact(x)
        TopRounded(x,70);

    Roof(x) ->
        CreateFact(x)
        Rounded(x,60);

    Walls(x,p) -> CreateFact(x);
```

There are several kinds of rules:
- Construction rules, which create facts and, possibly, hierarchically decompose the current fact. For example House(x).
- Intra-scene rules, which describe relations concerning the current scene. For example HigherThanWide(x).
- Inter-scene rules, which describe relations between two scenes. For example PutOn(x1,x2).
- Form rules, which describe the form of the current scene. For example Rounded(x,60).

2.3.2 Set of arithmetic or geometric constraints

The external description given by the designer is translated into a Constraint Satisfaction problem, that is a set constraints on variables taking their values in their corresponding domains.

This kind of internal description model is used by the DE2MONS declarative modeler, a modeler whose main properties are:
• A multi modal interface,
• A generation engine limited to the placement of objects,

• A constraint solver able to process dynamic and hierarchical constraints.

The modeler uses a multi modal interface allowing descriptions by means of the voice, the keyboard (natural language), a data glove or 3D captors informing the system of the user's position. This external description is translated in an internal description model made of linear constraints.

MultiFormes is also using, in its newer versions, an internal description model made of a CSP. Each CSP is a set of arithmetic constraints on finite domains. A special form of primitive arithmetic constraints, CLP(FD) is used to improve the scene generation process. The CLP(FD) form is based on the use of only primitive constraints of the form X **in** r, where X is a variable taking its values in a finite domain and r is a range. More information on CLP(FD) may be found in [14, 19, 20, 34, 38, 45] and in the first chapter of this volume.

The last version of MultiFormes uses an internal description model made of geometric constraints. A *geometric CSP* is defined as a triplet (V, D, C) where V is a set of variable points (or vertices), D is a set of 3D sub spaces, domains of the points of the set V, and C is a set of geometric constraints on points. A solution of a geometric CSP is a set of positions in the 3D space verifying all the constraints and belonging to the domain of each variable point.

For reasons of compatibility with the arithmetic constraint solver of MultiFormes, constraints of a geometric CSP are binary constraints of the form X **in** S, where X is a variable point and S a restriction of the 3D space.

All the high level properties used by the modeler have to be translated to a set of binary geometric constraints. For instance, the property "The three points A, B and C are aligned" will be translated to the set of binary geometric constraints:
A **in** (BC),
B **in** (AC),
C **in** (AB).

The VoluFormes [21] declarative scene modeler also uses an internal description model made of a set of constraints. The external high level description uses quasi-natural language to describe relative placement of bounding boxes, as well as properties of each bounding box. For example, properties like *small, medium, bigger than, in front of,* etc. may be used in this external description. This description is translated into a set of constraints, not necessarily linear.

The produced constraints use 8 parameters to describe a box. These parameters are:

T: the size, that is the bigger dimension of the box;

a: the ratio W/L, where W is the width and L the length of the box;

b: the ratio H/W, where H is the height of the box;

Cx: the x-coordinate of the center of the box;

Cy: the y-coordinate of the center of the box;

Cz: the z-coordinate of the center of the box;

V: this parameter indicates whether the box is horizontal or vertical;

S: this parameter indicates the direction of the box around its big axis.

It is presumed that $H \leq W \leq L$. Thus, $T = L$ and $a, b \leq 1$. If T, a and b are known, the dimensions of the box are computed by the formulas:

$L = T$;

$W = T.a$;

$H = T.a.b$;

2.4 The generation phase in declarative scene modeling

The internal description model obtained in description phase of declarative scene modeling, is used to generate solutions during the generation phase. As descriptions are generally imprecise in declarative scene modeling, more than one solutions are often generated from a single internal description model.

The used techniques to generate solutions from an internal description model depend on the kind of internal model. As we have already seen above, the main internal description models used in declarative scene modeling are the *Set of Rules and Facts* and the *Constraint Satisfaction Problem* (CSP) model.

2.4.1 Set of rules and facts

The PolyFormes declarative scene modeler uses an inference engine for scene generation from its internal description model, made of rules and facts. The inference engine generates all regular or semi-regular polygons corresponding to this description.

The inference engine of PolyFormes works according to a forward chaining mechanism. Processed rules are chosen using pattern matching, based on unification of facts of the heading of a rule with fact of the base of facts. Facts produced by application of rules are added to the base of facts, if they are new ones. The work of the inference engine stops when the base of facts is saturated (no new facts are produced) or when a contradiction occurs. In the first case, a solution, modeled as a list of polygons, is obtained. All possible solutions corresponding to a given, (precise or imprecise) description may be given by PolyFormes.

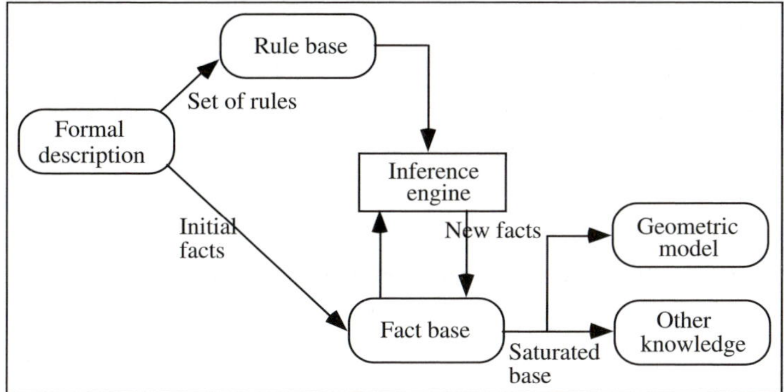

Fig. 2.2: The general working mechanism of PolyFormes

Figure 2.2 shows the general working mechanism of PolyFormes, while Fig. 2.3 shows an example of polyhedron generated by the PolyFormes declarative modeler.

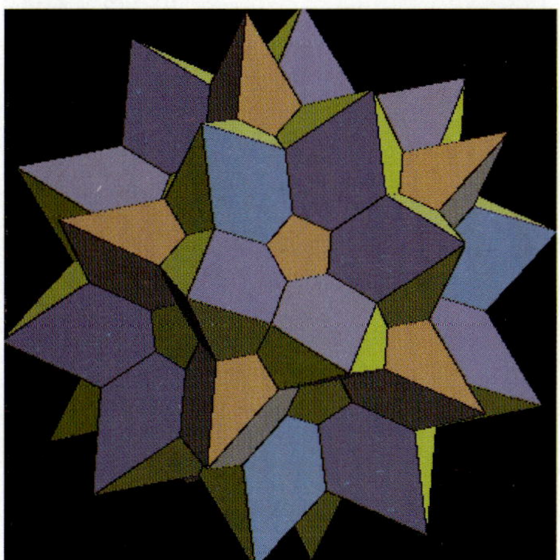

Fig. 2.3: Example of polyhedron generated by PolyFormes (Image P. and D. Martin)

The first version of the MultiFormes declarative scene modeler also used a set of rules and facts as internal description model.

At the beginning of the processing, the only fact in the base of facts is the workspace, defined by coordinates of its starting (lower-left-front) point and its width, height and depth. From this initial fact, other facts are created and will constitute a tree, corresponding to a solution generated by the modeler. The inference engine, applied to the rules, may generate all possible solutions for a given internal description model. Each fact represents a sub-scene of the final scene and is defined by its starting point, its width, height and depth, and, sometimes, by its form, expressed by a form rule attached to it.

In a general manner, the scene generation uses the hierarchical multi-level generation technique, which can be described as follows:

Starting with an approximate sketch of the scene, the modeler applies description rules in order to refine progressively this approximate representation and to get, finally, a set of solutions (scenes) verifying the properties wanted by the user.

The generated scenes generated are solids or a union of solids. Generation of other kinds of scenes by top-down multi-level description, used by the internal description model, and hierarchical generation is possible but not experimented.

In order always to get a finite number of solutions, even with imprecise properties, a discrete workspace is used, defined during the description step. The use of a discrete workspace is a tool of limitation of the number of generated scenes, which also permits, in many cases, to insure oneness of the proposed solutions.

The main advantages of the hierarchical multi-level generation technique are:
* generation can be made at various detail levels; thus, one can stop the scene generation at a given detail level, even if the scene description includes additional detail levels.
* generation insures inheritance of properties ; thus, if one specifies that *House* is pasted on the left of *Garage*, this property will be true for all sub-scenes of *House*, at any level. On the other hand, if *Residence* has the property *Rounded*, this property is inherited by the sub-scenes of *Residence* in order to form a scene with rounded aspect.

The work of the hierarchical multi-level generation technique is insured by a special inference engine which processes the rules and facts of the internal model. In the following paragraphs we will explain how the inference engine and the main already defined rules function.

The inference engine

The inference engine is a special one, which uses backward resolution and breadth-first search. Starting from a goal to state, it scans the rules of the internal model and, if one of the rules is applicable, it applies the rule as long as the rule remains applicable.

A rule is applicable if the two following conditions are verified together:
- its conclusion can be unified with the current goal,
- the base of facts generated by this rule is not saturated.

The nature of a rule processed by the inference engine can be one of the following :
- non-terminal rule,
- terminal rule,
- the special rule *CreateFact*.

Only non-terminal rules are applicable many times.

The deduction rules

Non-terminal rules are automatically generated by the modeler from the properties specified by the user. They use, in their conditions, terminal rules, already defined, or other non-terminal rules. The rule:

```
House(x) ->
      CreateFact(x)
      Habitation(y)
      Garage (z)
      PasteLeft (y,z);
is a non-terminal rule.
```

The principle of non-terminal rules processing by the inference engine is very simple:

If the rule can be unified with the current goal, unification is performed and the current goal is replaced by a set of new goals. For each of these new goals, the inference engine will try to unify it with one or many rules.

Terminal rules of the internal model are rules defined by the expert with no reference to other rules. Top rounded, Put on, etc. are terminal rules. They are applied once, if they can be unified with the current goal. They work by using the discrete workspace saved by the modeler like a fact of the internal model. Terminal rules can be precise or imprecise. The special rule *CreateFact* will be included among terminal rules, although it cannot be imprecise.

(1)   The rule CreateFact.

Every time it is applied, this rule creates a fact, describing an occurrence of the current node of the hierarchical description and containing information about this occurrence, like: the bounding box of the corresponding scene, a representation of the scene at the current detail level by a set of control points, links with parent and children of the occurrence. This information will be completed later, by application of other rules. The application of the rule *CreateFact* has for consequence the generation of facts like the one of Fig. 2.4.

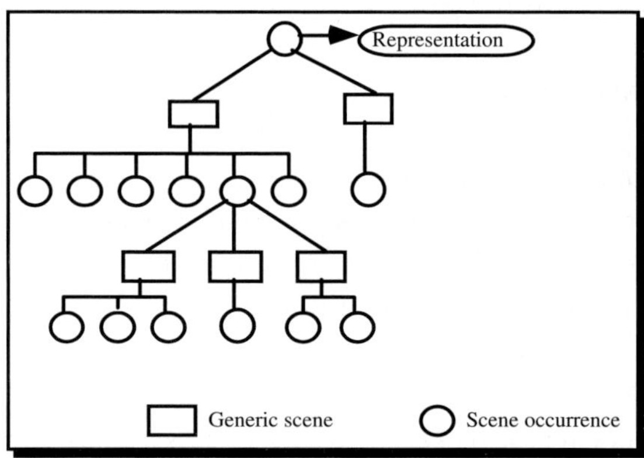

Fig. 2.4: Fact generated by MultiFormes

At the moment of generation of a scene or sub-scene occurrence, a representation is attributed to this occurrence, the one of its embedding box. This initial representation will be refined, during the generation step, by rules which describe form properties of the current scene or, by inheritance, of its ancestors.

When the fact representing one solution in the root level is completely defined, a module of solution extracting is called by the inference engine. This module extracts from the fact of Fig. 2.4 a set of trees, each of which represents a solution (scene).

(2)   Size inter-dimensions rules.

These rules are applied to all the occurrences of a generic scene or sub-scene. Each time they are applied to an occurrence, they also operate on all occurrences of all descending nodes of the current occurrence.

It is very important for these rules to be applied as soon as possible, in order to avoid generation of descending nodes for occurrences which do not verify properties described by size inter-dimensions rules.

(3)   Size inter-scene rules.

These rules are applied to all the occurrences of two sub-scenes of a generic scene or sub-scene. Application of a size inter-scene rule has for consequence the suppression of some occurrences of sub-scenes which do not verify the corresponding size property.

For cost reduction reasons, these rules should be applied after size inter-dimensions rules and before placement rules.

(4)   Placement rules.

These rules are applied to all the occurrences of two sub-scenes of a generic scene or sub-scene or to all the occurrences of a generic scene or sub-scene. They can be simple placement (*Garage* is placed at (0, 0, 1)) or inter-scene (*Roof* is put on *Walls*) placement rules.

Application of a placement rule brings suppression of some occurrences of sub-scenes, generated by the rule *CreateFact*, which do not verify the corresponding size property.

Sub-scene occurrences of a scene occurrence, verifying the placement properties, are put together into a list of n-uples, called **valid n-uples**.

(5)   Form rules.

These rules are applied to all the occurrences of a generic scene or sub-scene and can be precise (*Roof* is 80% rounded) or imprecise (*Roof* is almost rounded).

Each time a form rule is applied to an occurrence of a generic scene or sub-scene it may also operate, by inheritance, on all occurrences of all descending nodes of the current occurrence.

We have seen that a representation is attributed to each scene or sub-scene when it is initially created. This representation is the one of its bounding box, materialized by 6 times 16 control points. Fig. 2.5 represents bounding boxes of *Roof*, *Walls* and *Garage* for an occurrence of *House*.

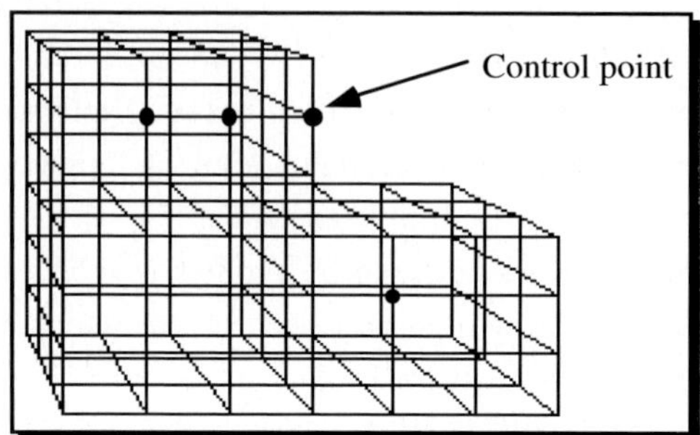

Fig. 2.5: Bounding boxes of Roof, Walls and Garage

Application of a form rule operates on control points of the scene representation and brings a deformation of the scene (Fig. 2.6).

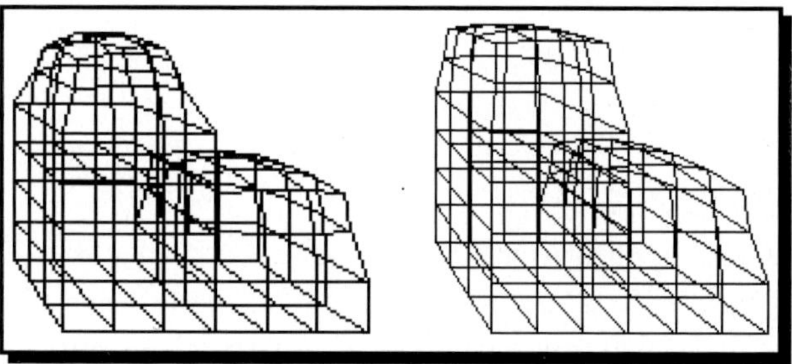

Fig. 2.6: Two occurrences of the whole scene after deformation

Form rules are time consuming rules; thus it is important to do not apply these rules to scenes which could be suppressed later. Therefore, these rules are applied at last.

(6)   Imprecise rules.

All terminal rules, except rule *CreateFact*, can be precise or imprecise. Imprecise rules work in different ways, according to their nature.

Thus, for size inter-dimension, size inter-scene and placement rules, application of an imprecise rule has for consequence the suppression of all the occurrences generated by the rule *CreateFact* which do not verify the corresponding property and the suppression of their descending nodes. These kinds of rules perform a posteriori suppression whenever they fail.

In return, imprecise form rules never fail if the other rules in the description are successful. They generate many solutions, obtained by deformation, from the scene verifying the other properties. Thus, application of the rule *Roof* **is almost rounded** could involve generation of some representations of the current scene, corresponding respectively to 70% Rounded, ..., 95% Rounded.

2.4.2 Set of arithmetic or geometric constraints

Declarative modelers using an internal description model made of a set of arithmetic or geometric constraints have to resolve a CSP. CSP resolution methods are different, according to the particular declarative scene modeller.

The **DE2MONS** declarative scene modeler uses a resolution method based on the notion of *hierarchical constraint*, that is constraints with priorities, assigned by the user.

In order to generate solutions from its internal description model, DE2MONS uses a linear constraint solver, ORANOS, able to process dynamic constraints (new constraints can be added during generation) and hierarchical constraints. Whenever there is no solution for a given description, constraints with low priority are released in order to always get a solution. The solver computes only one solution for a given description.

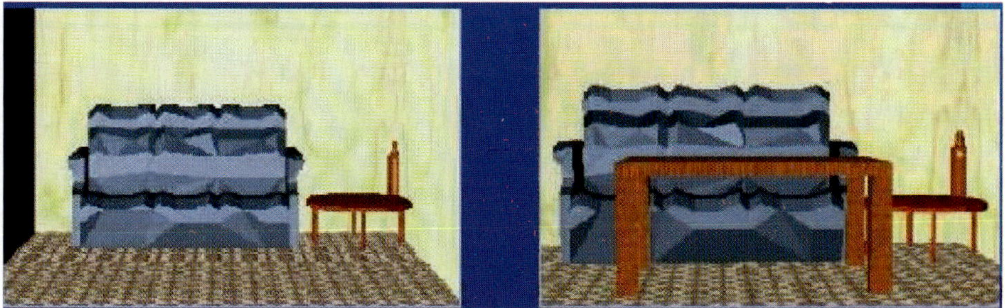

Fig. 2.7: Furniture pieces arrangement in a room with DE2MONS

DE2MONS and its constraint solver only manage placement of already existing objects in a given space. In Fig. 2.7 one can see a furniture placement in a room, obtained by the DE2MONS modeler.

The **VoluFormes** scene modeler uses linear and parabolic constraints to manage the placement of boxes in the space. As we have seen above, each box is defined by 8 parameters and a set of constraints. Among these parameters, 6 may change their values. So, the problem is to find a value for the 6-ple (T, a, b, Cx, Cy, Cz) satisfying the corresponding constraints.

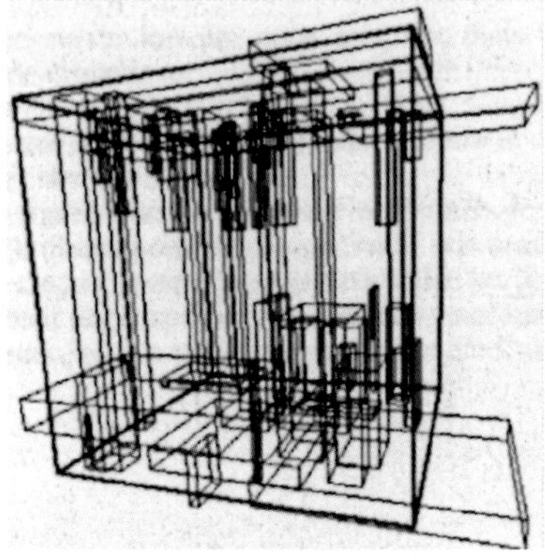

Fig. 2.8: Boxes arrangement with VoluFormes

The resolution method is based on a combination of interval reduction, obtained from the constraints associated to a box, and random choice of the values of parameters. The boxes are created in an incremental manner. Cases where no solution is found are resolved by compromise. If a solution is not satisfactory, another random choice may be done. This resolution method does not allow to completely explore the solution space. Fig. 2.8 presents a boxes arrangement obtained with VoluFormes.

Fig. 2.9: Scene obtained with Voluformes by form growth mechanisms

Production of a set of boxes is the first step of the whole work of Voluformes. In the second and last step, the designer creates scenes by putting germs in the boxes and selecting the growth mechanism, among a reduced number of predefined imperative mechanisms. In Fig. 2.9 one can see an example of scene obtained by form growth.

We have seen above that the **MultiFormes** declarative scene modeler uses an internal description model made of arithmetic or geometric constraints. The used resolution techniques of these CSP follow classical CSP general resolution methods but they are different in the details, especially in the used heuristics in order to get efficient resolution. The used methods in the case of arithmetic constraints, as well as in the case of geometric constraints, will be detailed below.

Arithmetic constraint satisfaction

As each constraint has to be satisfied, the problem of the modeler is to find (at least) a set of values, one per variable, satisfying all the constraints. In other words, the modeler has to obtain solutions for a *constraint satisfaction problem* (CSP). We have seen above that, in order to avoid some drawbacks of general form CSP and to obtain improvements due to the hierarchical decomposition, a new kind of primitive constraints and a new method to enforce problem reduction is used: constraint logic programming on finite domain, CLP(FD).

The resolution process for a CSP is an iterative process, each step being composed of two phases: the enumeration phase and the propagation phase.

1. The *enumeration phase* is made of two parts:
    *   Selection of a variable;
    *   Selection of a value for this variable.

    It is possible to use various orders in selecting variables and values. Some selection orders give best results than others. The variable selection order giving the best results it the order which follows the hierarchical decomposition tree given in the description phase.

2. The *propagation phase* takes into account the value of the variable chosen in the enumeration phase and tries to reduce the constraint satisfaction problem

by propagating this value. Propagation of the value of the selected variable can be made in different manners:

- Propagation through the affected constraints of the already instantiated variables. In this method, called "look-back", the resolution engine checks if the new value is compatible with the previously chosen values of variables.
- Propagation through all variables, even the not yet instantiated. This kind of propagation is called "look-ahead" and can avoid future conflicts.

If the "look-ahead" technique performs arc-consistency tests only between an instantiated and a not yet instantiated variable, it is called "forward-checking". If arc-consistency tests can be performed between not yet instantiated variables that have no direct connection with instantiated variables, in this case, the method is called "partial look-ahead".

The CLP(FD) method [19, 20] is based on the use of only primitive constraints of the form X **in** r, where X is a variable taking its values in a finite domain and r is a range. A range is a dynamic expression of the current domain of a variable and uses references to other variables via four *indexicals*, **min** (the min of the domain of the variable), **max** (the max of the domain of the variable), **dom** (the domain of the variable) and **val** (the value of the variable). Arithmetic expressions using indexicals, integer values and arithmetic operators can be used to describe a range.

In the final set of primitive constraints to be processed, each variable appears only once under the form: X **in** r1&r2&...&rn, where r1, r2, ..., rn are ranges and & denotes the union of ranges. This final form may be automatically obtained by applying rewriting rules. For instance, constraint "X=Y+Z" will give the set of primitive constraints:

X **in** min(Y)+min(Z) .. max(Y)+max(Z)
Y **in** min(X)-max(Z) .. max(X)-min(Z)
Z **in** min(X)-max(Y) .. max(X)-min(Y)

During the resolution process, a value is given to a variable X in order to reduce its range to a single value. The solver is then looking for indexicals using X in the domains of all the other variables and updates them. The ranges of the variables using these indexicals is reduced and this reduction is propagated to other variables. If the range of a variable becomes empty, there are no solutions for the chosen value of X and another value has to be tried.

With primitive constraints, various propagation strategies can be applied, depending on indexicals taken into account. Thus:

- Full look-ahead is obtained by using the indexical **dom**,
- Partial look-ahead is obtained by using only indexicals **min** and **max**. In this way, holes are not transmitted.
- Forward checking by using the indexical **val**, delaying the check of the constraint until the involved variable is ground.

Hierarchical Decomposition-based Improvements

The CLP(FD) method can be more or less efficient, depending on the order to handle variables. The hierarchical decomposition tree used by DMHD gives the possibility to improve this order [14, 34, 38].

Thus, the hierarchical decomposition tree can be processed by levels, the nodes of each level being rearranged in order to first visit simpler sub-trees, that is, sub-trees of less depth (Fig. 2.10). Exploration by levels and reordering permit to apply the "fail first" heuristic which tries to reach a possible failure as soon as possible.

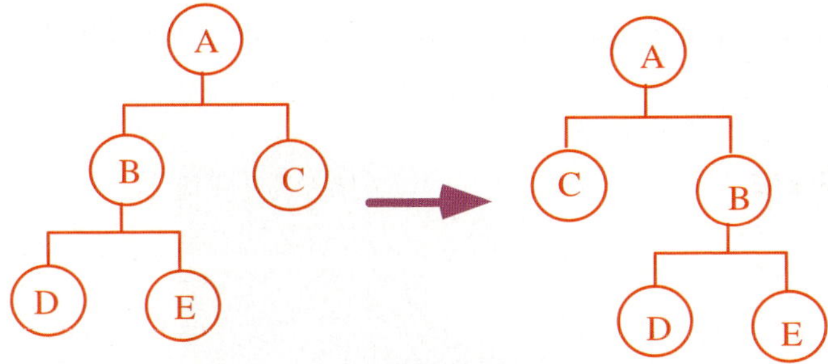

Fig. 2.10: Reordering according to the depth of sub-trees

Another possibility to improve exploration by levels is to order the variables associated to a node of the hierarchical decomposition tree according to the number of constraints they share. This heuristic allows to process first, variables sharing a big number of constraints to reach sooner a possible failure.

A last possibility to improve resolution is to improve backtracking whenever a failure is reached. Let us consider the case of the tree of Fig. 2.11. During the processing of the level of nodes D, E, F and G, when a failure occurs in nodes E, F or G, backtracking is classically performed to the previous node (D, E or F respectively).

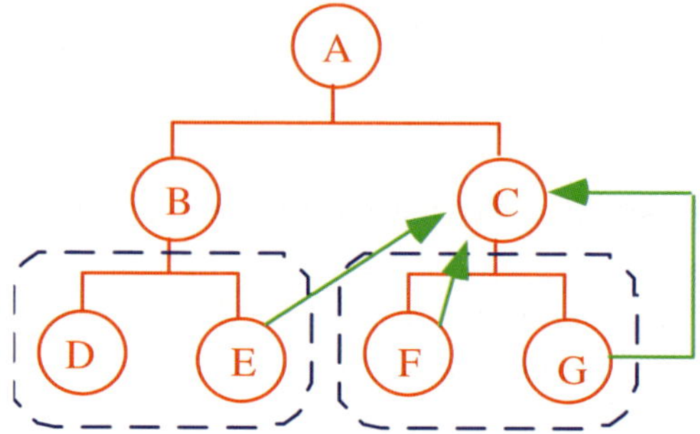

Fig. 2.11: Grouping of nodes and intelligent backtracking

However, values of variables of nodes D and E are completely independent from values of variables of nodes F and G. If a failure occurs in node F or G, nodes D and E are not responsible for this failure. With intelligent backtracking, if a failure occurs in nodes F or G, backtracking is performed to the last node of the previous level. So, for the example of Fig. 2.11, backtracking is performed to the node C, without exploring all the remaining possibilities for nodes D and E.

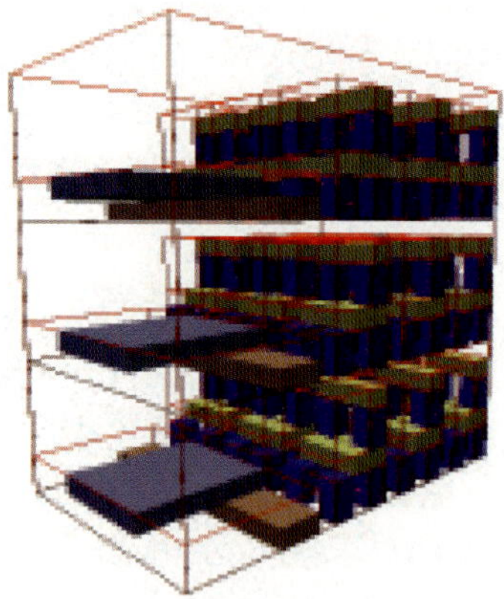

Fig. 2.12: Inside a 3-floor building (Image P.-F. Bonnefoi)

Intelligent backtracking is based on independence of various parts of the CSP and these parts are easy to determine with the hierarchical decomposition tree.

This independence has been exploited to make parallel the resolution process in MultiFormes. Fig. 2.12 shows an example of scene generated with the arithmetic CSP solver of MultiFormes.

Geometric constraint satisfaction

The resolution process is based on a classical chronological backtrack algorithm. Variable points are sequentially processed. The algorithm tries to find a value in the domain of the current variable point, satisfying the constraints with the current values of already processed variable points. In case of failure, another value of the current variable point is reached. In case of failure for any possible value of the current variable point, backtracking is performed to the last processed variable point and a new value is tried for this point.

The domain of a variable point is computed by computing the intersection of all the domains assigned to the variable point by properties including this point. If the final domain is empty, there is no solution for this variable point, that is there no position for this point satisfying all its constraints. If the obtained final domain is continuous, it has to be sampled.

The resolution process uses intersection computation of geometric objects as spheres, boxes, straight lines etc. The intersection computation process may produce either a discrete or a continuous domain. When a discrete domain is obtained, all the possible discrete values have to be tried by the solution searching process. Otherwise, the obtained continuous domain must be sampled in order to get a finite number of solutions.

The geometric entities used by the solver are: point, straight line, half-straight line, segment, plan, convex polygon, box, sphere, circle, arc of circle and ball (solid sphere) [35, 38]. However, the intersection of some geometric entities may produce an entity not present in the above list. In such a case approximation is used in order to get a known kind of geometric entity. The problem is that the sampling process, applied to an approximated domain, may produce points which are not solutions because they don't belong to the real domain. Thus, a filtering step has been added in the resolution process, in order to verify that a chosen value belongs to the real domain. In Fig. 2.13, one can see illustration of this problem for the case of intersection of a circle and a convex polygon.

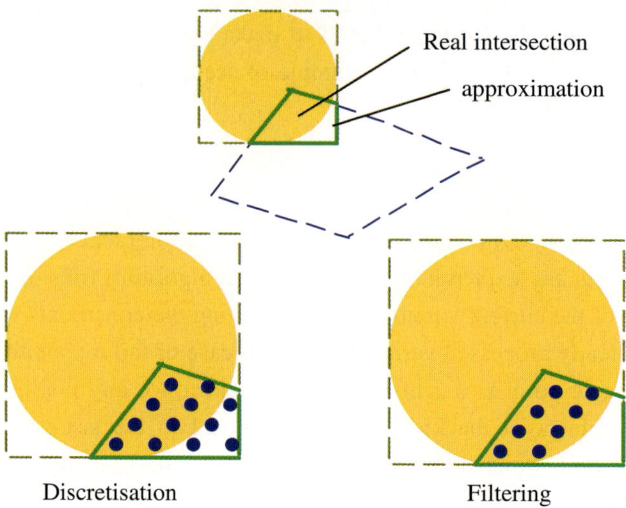

Fig. 2.13: Approximation, sampling and filtering of an intersection

Approximation and sampling of intersections have to be delayed as late as possible in order to avoid problems of lack of solutions which may occur if an approximated sampled domain is intersected by another approximated sampled domain. A special mechanism of the geometric CSP solver rearranges intersection computations in order to use approximation and sampling only at the end of the intersection computation process.

The following two problems may affect the efficiency of the geometric CSP solver.

The first one is due to the need of sampling. Existence of solutions may depend on the instantiation order of the variable points. For example, let us consider the case where two points C and D have to belong to the already instantiated segment (AB) and BD has to be 4 times longer than AC. If the sampling mechanism gives three possible positions on the segment (AB), the following cases may occur:

- The variable point C is instantiated before the point D. None of the three possible positions of C is a solution because, the constraint BD = 4 * AC is not satisfied for any position of D (Fig. 2.14, left).

- The variable point D is instantiated before the point C. It is possible to find a solution for this problem (Fig. 2.14, right).

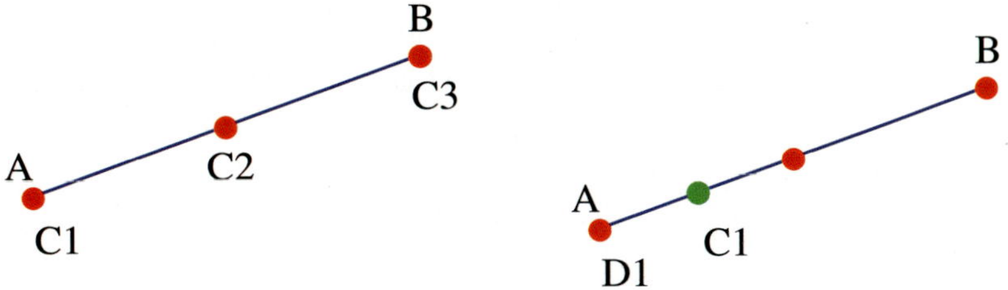

Fig. 2.14: The instantiation order problem. If C is instantiated before D, the solver doesn't find solution, even if a solution exists.

Another problem is the one of missing constraints. For the user, transitivity is an obvious property. So, if he (she) has to tell the modeler that segments (AB), (CD) and (EF) must have the same length, he (she) may think that it is enough to tell that (AB) = (EF) and (CD) = (EF). In fact, a constraint is missing, because the modeler does not know transitivity. In order to complete the description, a term rewriting system based on the Knuth-Bendix algorithm has been added to the solver.

Fig. 2.15: A scene generated by the geometric constraint-based MultiFormes declarative modeler: a table and 4 chairs (Image W. Ruchaud)

Fig. 2.15 shows a scene obtained by the geometric CSP solver of the Multi-Formes declarative scene modeler. One can see reusing of scene elements described only once.

2.4.3 Further generation improvements

The hierarchical decomposition structure of Declarative Modeling by Hierarchical Decomposition (DMHD), used by MultiFormes, allows important improvements in generation phase, drastically reducing the processing time. However, if the initial description is very imprecise, many generated scenes are uninteresting because they are not close to the mental image of the designer. The problem is that the modeler cannot know the semantics the designer assigns to each used imprecise property. In order to help the modeler generate scenes close to the designers desires, machine-learning techniques have been associated to the scene generation engine of MultiFormes.

Two machine-learning techniques have been implemented with different versions of MultiFormes. The first one uses a neural network to learn the designers preferences, whereas the second technique uses genetic algorithms. Other kinds of machine-learning for declarative modeling were proposed in [23, 40].

2.4.3.1 Neural network-based machine-learning

In order to improve scene generation, an interactive machine learning mechanism has been implemented, based on the use of neural networks [8, 9, 10, 11, 13], applied to DMHD. This mechanism is used by the modeler in the following manner:

- During a learning phase, some scenes, generated by the modeler from the initial description, are selected by the user to serve as examples of desired scenes. Each time a new example is presented, the modeler learns more on the user's preferences and this is materialized by a modification of the values of weights associated to the connections of the network. At the end of the learning phase, an acceptation interval is calculated and assigned to each decision cell.
- After the end of the learning phase, the modeler is in the normal working phase : the weights of connections calculated during the learning phase are used as filters allowing the choice of scenes which will be presented to the user.

The used machine learning mechanism takes into account only relative dimension and position properties of a scene. Form properties are processed by the scene generation engine after selection of dimensions and positions of the bounding boxes of each sub-scene.

The neural network is created dynamically from the description of the scene and its structure is described in the following lines.

To each node of the scene description tree, are assigned:

*   A network whose the input layer is composed of two groups of neurons (see Fig. 2.16) : a group of two neurons whose inputs are w/h and w/d where w, h and d are respectively the width, the height and the depth of the scene associated with the current node; a group of neurons whose inputs are the results of acceptance of the other nodes of the scene. The network contains another layer of two neurons which work in the following manner: the first one evaluates the quantity i1*w1+ i2*w2, where wk represents the weight attributed to the connection between the k-th neuron of the input layer and the intermediate layer and ik the k-th input; it returns 0 or 1 according to whether this weighted sum belongs to a given interval or not. The values of weights wk can be modified during the learning phase. The second neuron computes the sum of acceptance results coming from the second group of neurons of the input layer. Its output function returns 0 or 1 according to whether this sum is equal or not to the number of acceptance neurons. The decision layer of the network contains one neuron and computes the sum of the outputs of neurons, returning 1 (acceptance) or 0 (refusal) according to whether this sum is equal to 2 or not.

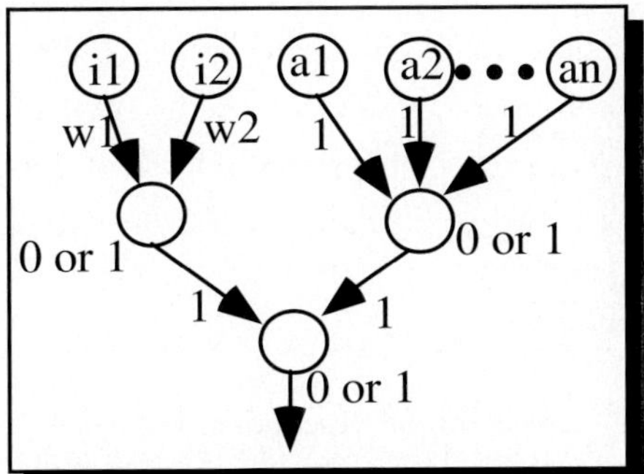

Fig. 2.16 : local neural network

*   n*(n-1)/2 networks, corresponding to all possible arrangements of two sub-scenes of the current scene, where n is the number of child-nodes of the current node. These networks have the same structure and the same behavior as the others, excepting that the inputs of the first group of neurons of the input layer are dx/dy and dx/dz, where dx, dy and dz are the

components of the distance d between the two sub-scenes. The values of weights wk can be modified during the learning process.

Let us consider the scene House, described by the following Prolog-like pseudo-code:

```
House(x) ->
      CreateFact(x)
      Habitation(x1)
      Garage(x2)
      PasteLeft(x1,x2);

Habitation(x) ->
      CreateFact(x)
      HigherThanWide(x)
      HigherThanDeep(x)
      Roof(x1)
      Walls(x2)
      PutOn(x1,x2);

Garage(x) ->
      CreateFact(x)
      TopRounded(x,70);

Roof(x) ->
      CreateFact(x)
      Rounded(x,60);

Walls(x,p) -> CreateFact(x);
```

The generated neural network for this description is shown in Fig. 2.17.

The acceptance process of a scene generated by the modeler is the following:
- If the current node of the generated scene is accepted at the current level of detail and relative placements of all the sub-scenes of the node at the next level of detail are also accepted by the associated neural networks, the scene is partially accepted at the current level of detail and acceptance test is performed with each child-node of the current node. A scene is accepted at the current level of detail if it is partially accepted for each node up to this level of detail.

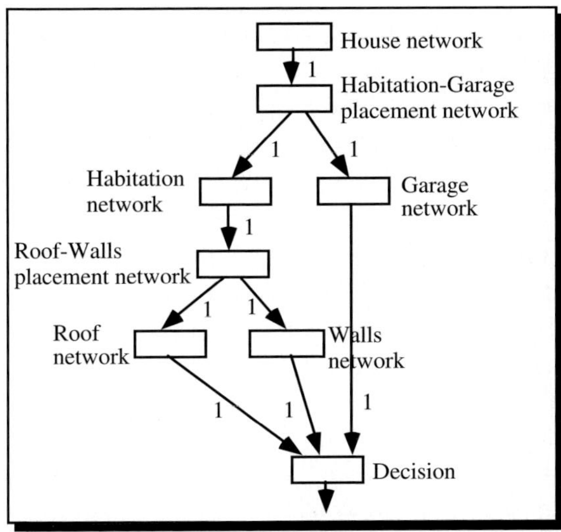

Fig. 2.17: Neural network generated for the "House" scene description

- A scene generated by the modeler is accepted if it is accepted at each level of detail. Otherwise, the scene is refused and it is not presented to the user.

The machine learning process

For each neural network of the general network, machine learning is performed in the same manner.

From each new example of scene presented by the user among the scenes generated by the modeler, each local network learns more and more and this fact is materialized by modifying the weights of connections between the association (intermediate) layer and the decision (output) layer. The formula used to adjust a weight w is the following:

$$wi+1 = w0 - wi\ V(Xi+1)$$

where

wi represents the value of the weight w at step i,

w0 represents the initial value of the weight w,

Xi represents the value of the input X at step i,

V(Xi) represents the variance of values X1, ..., Xi.

This formula permits to increase, or not to decrease, the value of the weight when the corresponding input value is constant and to decrease the value of the weight when the corresponding input value varies.

When, at the end of the learning phase, the final value of the weight w has been computed, the value of the quantity (w - w0) / m is estimated, where m is the number of examples selected by the user. If the value of this quantity is less than a threshold value, the weight w takes the value 0; otherwise, it takes the value 1.

The quantity computed by the output function of the first neuron of the intermediate layer of each local network is:

S = w1 Xi + w2 Yi,

where Xi and Yi are the input values at step i.

Thus, at the end of the machine learning phase, each output function of the first neuron of the intermediate layer of each local network will compute the quantity:

Sa = w1 A(X) + w2 A(Y)

where A(X) represents the average value of the input values X1, ..., Xn.

During the phase of normal working of the modeler, a solution is partially accepted by a local neural network if the output value computed by the output function of the first neuron of the intermediate layer belongs to the neighboring of the value Sa.

Let us consider three cases for each local network:

1.   w1 = w2 = 0.
The value of Sa is then equal to 0 and all solutions are accepted.

2.   w1 = 0, w2 = 1.
The value of Sa is then equal to A(Y). Only the input value associated with w2 is important and only scenes which have, for their corresponding to the current local network part, input values close to the value of A(Y) are selected.

3.   w1 = w2 = 1.
The value of Sa is then equal to A(X) + A(Y). The two quantities associated with w1 and w2, are important. The parts of scenes selected by the current local neural network will be the ones whose input values associated with w1 are close to A(X) and input values associated with w2 are close to A(Y), but also other scenes for which the sum of entry values is equal to A(X) + A(Y) whereas they should not be accepted.

Discussion

Although neural networks are fully efficient for linearly separable problems, their application to declarative modeling for a non linearly separable problem, selecting scenes close to those wished by the designer, gives results globally satisfactory for exploration mode generation, because it lightens the designer's work by filtering the major part of non interesting solutions. Machine learning is performed with little information and already learnt knowledge can be used for continuous machine learning where the modeler avoids more and more non interesting scenes.

Fig. 2.18 shows some rejected scenes during the machine learning phase for the above "House" scene description. The purpose of this phase was to learn that the "Habitation" part of the scene must be wider than the garage part. In Fig. 2.19, one can see some generated scenes for the "House" scene description, after the machine learning phase.

Of course, this kind of interactive machine learning reduces the number of solutions shown to the user but not the number of tries; the total exploration time remains unchanged. Another drawback of the method is that learning is not efficient if the number of given examples is small and that learning can be avoided if the given examples are not well chosen.

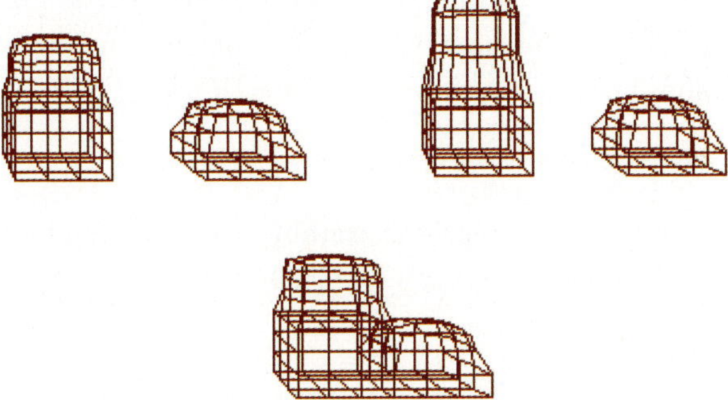

Fig. 2.18: Scenes generated from the "House" description and rejected by the user.

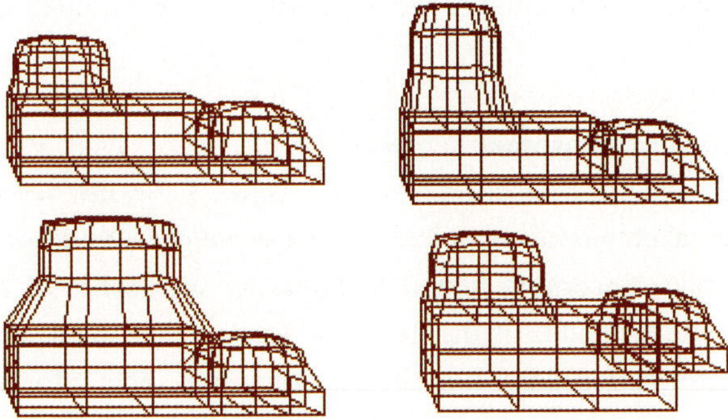

Fig. 2.19: Machine learning for the "House" description. Only generated scenes with the wished shape are presented to the user.

We think that it is possible to use the same neural network to reduce the number of tries during generation. To do this, after the machine learning phase, the neural network has to be activated as soon as possible, after each enumeration phase in the constraint satisfaction process. In this manner, the next step of the resolution process will take place only if the scene goes through the neural network filter.

2.4.3.2 Genetic algorithm-based machine-learning

Genetic algorithms [37, 41, 45] can also be used to perform machine learning because they are supposed to follow natural evolution lows, based on selection of individuals (chromosomes). Chromosomes are selected, according to their importance, to participate to the next generation. Finally, an important number of generated chromosomes has the properties corresponding to the evaluation function.

Genetic algorithms based machine learning has been implemented on Multi-CAD [44, 36], an information system for CAD, which uses the MultiFormes' generation engine to generate scenes.

In this implementation, the initial generation is composed of an arbitrary chosen number of scenes generated by the generation engine of MumtiFormes and the purpose of machine learning is to generate other scenes verifying the properties wished by the user. Each initial scene is a chromosome, encoded in a special manner.

(1) Encoding chromosomes

Information taken into account to encode a scene generated by the MultiFormes engine is the one concerning bounding boxes of sub-scenes of the scene corresponding to terminal nodes of the generation tree. Each sub-scene is represented by pertinent information describing its bounding box. This information is the following:
- coordinates of the bottom-front-left corner of the bounding box,
- width, depth and height of the bounding box.

Assuming that each coordinate, as well as width, depth and height of a bounding box are integer numbers, each bounding box of sub-scene needs 6 integers to be described (Fig. 2.20).

x	y	z	w	d	h
0	1	3	5	3	6

Fig. 2.20: sub-scene representation

As genetic algorithms use chromosomes encoded as binary strings, each information describing a sub-scene is encoded as a binary string. In the current implementation, the length of each binary string is limited to 3, in order to represent quantities from 0 to 7. So, the terminal sub-scene of figure 9 will be encoded as shown in Fig. 2.21.

000	001	011	101	011	110

Fig. 2.21: binary string encoding for sub-scene of figure 9.

The whole scene is encoded with a chromosome including all encoded terminal sub-scenes of the scene.

(2) The genetic algorithm

The purpose of the genetic algorithm is to create new generations of chromosomes, i.e. solution scenes, from an initial population by applying genetic operations to the chromosomes of the current generation in order to get the next one.

The used genetic operations are the classical ones, that is *cloning, crossover* and *mutation*. The probability for a chromosome to be submitted to a genetic operation is proportional to the importance of this chromosome. The importance of a chromosome is determined interactively by the user who assigns a fitness score to each chromosome of a new generation.

The genetic algorithm starts with an initial set of solutions created by the generation engine of MultiCAD, which is the one of MultiFormes. The user assigns a fitness score to each solution and the genetic algorithm based scene generation process starts.

At each step of generation, chromosomes are selected according to their importance and the genetic operations are applied to them, in order to generate the next generation of solutions. Unlike in many genetic algorithm applications, mutation is applied to each step of the generation process. The whole process can be stopped when at least one of the solutions of the new generation satisfies the wishes of the user.

(3) Discussion

Fig. 2.22 shows two solutions generated by the implemented genetic algorithm, for the scene "House" described above. Solutions are visualized through a VRML file.

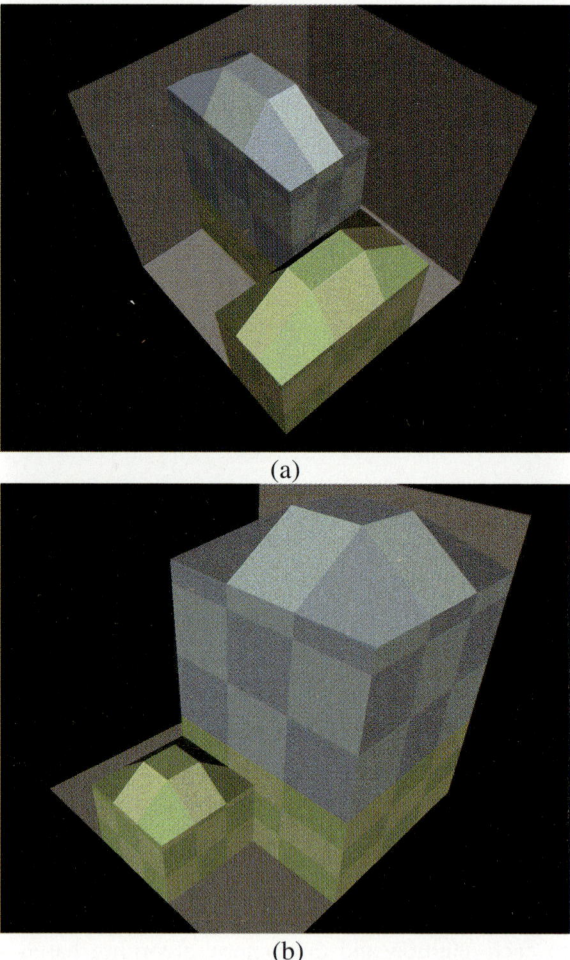

(a)

(b)

Fig. 2.22: scenes generated by the genetic algorithm for the scene "House"
(Images from TEI of Athens)

Genetic algorithm based learning is more suitable with the solution search generation mode. The obtained results are quite interesting and generated scenes are more and more close to the wishes of the user. The main advantage of the genetic algorithm approach is that only some solutions, the initial population, are generated using the time consuming constraint satisfaction techniques, used by the main generation engine of MultiFormes. The final solutions are obtained by the evolution process used in genetic algorithms, where the search is guided by the user who directs search on more promising parts of the search tree. An additional advantage is that the user has evaluate solutions by groups, rather than one by one as in the main generation engine of MultiFormes. This possibility permits a relative comparison of scenes and improves, generally, results.

On the other hand, as relative positions of the terminal sub-scenes of a scene are not encoded, it is possible to get undesired overlaps between these sub-scenes, especially during the first steps of the generation process. Moreover, some potentially interesting solutions may never appear, due to the choice of the initial population.

2.5   The scene understanding phase in declarative scene modeling

As declarative scene modeling generates several solutions and most of them can be unexpected, it is often necessary that the modeler offers scene understanding techniques in order to allow the designer to verify the properties of an obtained solution. Scene understanding can be visual or textual. Most of existing declarative modelers scene use simple scene display from an arbitrary chosen point of view. Very few declarative modelers use sophisticated scene understanding mechanisms.

PolyFormes uses a "matches-like" display mode allowing the user to better understand the shape of a generated polyhedron. In this kind of display, only the edges of the polyhedron are displayed but they are thickened (see Fig. 2.23).

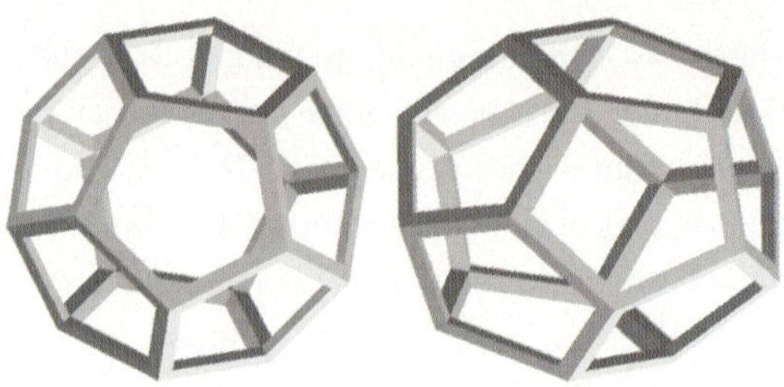

Fig. 2.23: "Matches-like" display of polyhedrons

MultiFormes uses more sophisticated techniques for scene understanding. These techniques use a good view criterion based on the scene's geometry and automatically compute a good point of view by *heuristic search* [25]. Some other techniques for computing a good view have been proposed for particular cases [26].

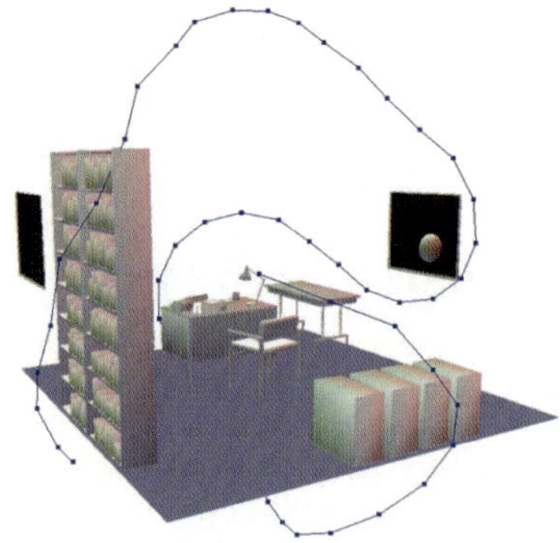

Fig. 2.24: Scene automated exploration by a virtual camera (Image G. Dorme)

As a single point of view is not always enough to understand complex scenes, MultiFormes also proposes an intelligent automatic scene exploration by a virtual camera, moving on the surface of a sphere surrounding the scene [27, 28, 29, .39]. An example of path of the virtual camera exploring a scene is shown in Fig. 2.24.

Neither DE2MONS nor VoluFormes propose specific scene understanding techniques. On the other hand, QuickVox, an intelligent scene modeler developed by Mathieu Kergall [33], uses a specific scene understanding technique. This modeler is not really a declarative one but it may propose more than one solution from a given description. It allows to design objects from their shadows on the three main planes defined by the three main axes of coordinates. The used scene understanding method proposes to split the produced scene, in order to detect cavities inside the scene. In Fig. 2.25 one can see an example of split scene designed with QuickVox.

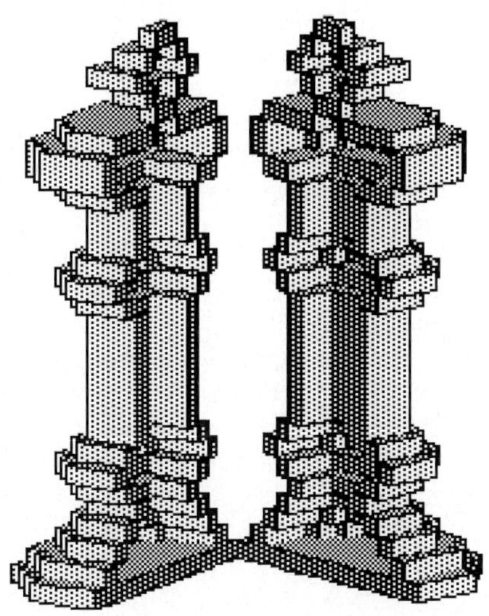

Fig. 2.25: Split view of a scene produced by QuickVox (Image M. Kergal)

2.6 Conclusion

In this chapter, we have tried to present a first application of Artificial Intelligence techniques in Computer Graphics, intelligent scene modeling. Intelligent scene modeling described in this chapter is mainly declarative modeling, (with one exception) presented from the angle of the used Artificial Intelligence techniques. Four declarative modelers have been chosen to illustrate the various manners to use Artificial Intelligence in scene modeling, in order to make easier the work of scene designer.

This chapter does not contain an exhaustive presentation of all research work on declarative modeling, even if, sometimes, a lot of details are given on used methods to find satisfactory solutions to open research problems. Our purpose was to show how Artificial Intelligence techniques may be used in each one of the main phases of intelligent scene modeling, that is *scene description, scene generation* and *scene understanding*. Lectors interested in declarative modeling may find more details and references in [45].

References

1. Lucas, M., Martin, D., Martin, P., Plemenos, D.: The ExploFormes project: some steps towards declarative modeling of forms. In: AFCET-GROPLAN Conference, Strasbourg, France, November 29 - December 1, vol. 67, pp. 35–49. BIGRE (1989) (in French); published in BIGRE
2. Plemenos, D.: A contribution to study and development of scene modeling, generation and display techniques - The MultiFormes project. Professorial Dissertation, Nantes, France (November 1991) (in French)
3. Plemenos, D.: Declarative modeling by hierarchical decomposition. The actual state of the MultiFormes project. In: International Conference GraphiCon 1995, St Petersbourg, Russia, July 3-7 (1995)
4. Plemenos, D., Tamine, K.: Increasing the efficiency of declarative modeling. Constraint evaluation for the hierarchical decomposition approach. In: International Conference WSCG 1997, Plzen, Czech Republic (February 1997)
5. Lucas, M., Desmontils, E.: Declarative modelers. Revue Internationale de CFAO et d'Infographie 10(6), 559–585 (1995) (in French)
6. Martin, D., Martin, P.: An expert system for polyhedra modeling. In: EUROGRAPHICS 1988, Nice, France (1988)
7. Martin, D., Martin, P.: PolyFormes: software for the declarative modeling of polyhedra. In: The Visual Computer, pp. 55–76 (1999)
8. Mc Culloch, W.S., Pitts, W.: A logical calculus of the ideas immanent in nervous activity. Bulletin of Mathematical Biophysics 5, 115–133 (1943)
9. Rosenblatt, F.: The perceptron: a perceiving and recognizing automaton. Project Para, Cornell Aeronautical Lab. Report 85-460-1 (1957)
10. Plemenos, D.: Techniques for implementing learning mechanisms in a hierarchical declarative modeler. Research report MSI 93 - 04, Limoges, France (December 1993) (in French)
11. Boughanem, M., Plemenos, D.: Learning techniques in declarative modeling by hierarchical decomposition. In: 3IA 1994 Conference, Limoges, France, April 6-7 (1994) (in French)
12. Plemenos, D., Ruchaud, W., Tamine, K.: New optimising techniques for declarative modeling by hierarchical decomposition. In: AFIG annual conference, December 3-5 (1997) (in French)
13. Plemenos, D., Ruchaud, W.: Interactive techniques for declarative modeling. In: International Conference 3IA 1998, Limoges, France, April 28-29 (1998)
14. Bonnefoi, P.-F.: Constraint satisfaction techniques for declarative modeling. Application to concurrent generation of scenes. PhD thesis, Limoges, France (June 1999) (in French)
15. Kochhar, S.: Cooperative Computer-Aided Design: a paradigm for automating the design and modeling of graphical objects. PhD thesis, Harvard University, Aiken Computation Laboratory, 33 Oxford Street, Cambridge, Mass. 02138, Available as TR-18-90 (1990)

16. Kochhar, S.: CCAD: A paradigm for human-computer cooperation in design. IEEE Computer Graphics and Applications (May 1994)
17. Sellinger, D., Plemenos, D.: Integrated geometric and declarative modeling using co-operative computer-aided design. In: 3IA 1996 conference, Limoges, France, April 3-4 (1996)
18. Sellinger, D., Plemenos, D.: Interactive Generative Geometric Modeling by Geometric to Declarative Representation Conversion. In: WSCG 1997 conference, Plzen, Czech Republic, February 10-14 (1997)
19. van Hentenryck, P.: Constraint satisfaction in logic programming. Logic Programming Series. MIT Press, Cambridge (1989)
20. Diaz, D.: A study of compiling techniques for logic languages for programming by constraints on finite domains: the clp(FD) system
21. Chauvat, D.: The VoluFormes Project: An example of declarative modeling with spatial control. PhD Thesis, Nantes (December 1994) (in French)
22. Poulet, F., Lucas, M.: Modeling megalithic sites. In: Eurographics 1996, Poitiers, France, pp. 279–288 (1996)
23. Champciaux, L.: Introduction of learning techniques in declarative modeling, PhD thesis, Nantes (June 1998) (France)
24. Kwaiter, G.: Declarative scene modeling: study and implementation of constraint solvers, PhD thesis, Toulouse, France (December 1998)
25. Plemenos, D., Benayada, M.: Intelligent display in scene modeling. New techniques to automatically compute good views. In: GraphiCon 1996, St Petersburg, Russia, July 1-5 (1996)
26. Colin, C.: Automatic computing of good views of a scene. In: MICAD 1990, Paris, France (February 1990)
27. Barral, P., Dorme, G., Plemenos, D.: Visual understanding of a scene by automatic movement of a camera. In: GraphiCon 1999, Moscow, Russia, August 26 - September 1 (1999)
28. Barral, P., Dorme, G., Plemenos, D.: Visual understanding of a scene by automatic movement of a camera. In: Eurographics 2000 (2000)
29. Barral, P., Dorme, G., Plemenos, D.: Intelligent scene exploration with a camera. In: International conference 3IA 2000, Limoges, France, March 3-4 (2000)
30. Jolivet, V., Plemenos, D., Poulingeas, P.: Declarative specification of ambiance in VRML landscapes. In: Bubak, M., van Albada, G.D., Sloot, P.M.A., Dongarra, J. (eds.) ICCS 2004. LNCS, vol. 3039, pp. 115–122. Springer, Heidelberg (2004)
31. Fribault, P.: Declarative modeling of livable spaces. Ph. D thesis, Limoges, France, November 13 (2003) (in French)
32. Vergne, J.: Declarative modeling of wood textures. MS thesis, Saint-Etienne, France (June 2003) (in French)
33. Kergall, M.: Scene modeling by three shadows. MS thesis, Rennes, France (September 1989) (in French)
34. Bonnefoi, P.-F., Plemenos, D.: Constraint satisfaction techniques for declarative scene modeling by hierarchical decomposition. In: 3IA 2000 International Conference, Limoges, France, May 3-4 (2002)
35. Ruchaud, W.: Study and realisation of a geometric constraint solver for declarative modeling. PhD Thesis, Limoges, France, November 15 (2001) (in French)

36. Vassilas, N., Miaoulis, G., Chronopoulos, D., Konstantinidis, E., Ravani, I., Makris, D., Plemenos, D.: MultiCAD GA: a system for the design of 3D forms based on genetic algorithms and human evaluation. In: SETN 2002 Conference, Thessaloniki, Greece (April 2002)
37. Goldberg, D.E.: Genetic Algorithms. Addison-Wesley, USA (1991)
38. Bonnefoi, P.F., Plemenos, D., Ruchaud, W.: Declarative modeling in computer graphics: current results and future issues. In: Bubak, M., van Albada, G.D., Sloot, P.M.A., Dongarra, J. (eds.) ICCS 2004. LNCS, vol. 3039, pp. 80–89. Springer, Heidelberg (2004)
39. Sokolov, D., Plemenos, D., Tamine, K.: Methods and data structures for virtual world exploration. The Visual Computer 22(7), 506–516 (2006)
40. Bardis, G.: Machine-learning and aid to decision in declarative scene modeling. Ph. D thesis, Limoges, France, June 26 (2006)
41. Makris, D.: Study and realisation of a declarative system for modeling and generation of styles by genetic algorithms. Application to architectural design. Ph. D thesis, Limoges, France, October 18 (2005)
42. Golfinopoulos, V.: Study and realisation of a knowledge-based system of reverse engineering for declarative scene modeling. Ph. D thesis, Limoges, France, June 21 (2006)
43. Dragonas, J.: Collaborative declarative modeling. Ph. D thesis, Limoges, France, June 26 (2006)
44. Miaoulis, G.: Study and realisation of an intelligent multimedia information system for computer-aided design. The MultiCAD project. Ph. D thesis. Limoges, France, February (2002)
45. Miaoulis, G., Plemenos, D. (eds.): Intelligent Scene Modelling Information Systems. Studies in Computational Intelligence, vol. 181. Springer, Heidelberg (2009)
46. Plemenos, D., Miaoulis, G., Vassilas, N.: Machine learning for a general purpose declarative scene modeller. In: International Conference GraphiCon 2002, Nizhny Novgorod, Russia, September 15-21 (2002)

3 Scene understanding

Abstract. A review of main scene understanding techniques is presented in this chapter. After a definition of the purpose and usefulness of scene understanding, visual scene understanding techniques will be presented, based on good view criteria. These techniques try to find the best (or one of the best) point of view for a given scene, in order to allow a good understanding of the scene by the user. The used scene understanding techniques will be divided in too categories: "a posteriori" and "a priori" techniques.

Keywords. Good point of view, Viewpoint complexity, Viewpoint entropy, Scene geometry, Visibility, Total scene curvature, Scene salience.

3.1 Introduction

Scene understanding is a very important area in Computer Graphics. At about the second half of the 80's, we have localized this problem which appeared to us to be very important. Very often, the rendering of a scene, obtained after a long computation time, was not possible to exploit because the choice of the point of view was bad. In such a case, the only possibility was to choose another point of view and to try again by running the time consuming rendering algorithm once again. We thought that it is very difficult to find a good point of view for a 3D scene when the working interface is a 2D screen. On the other hand, we thought that the choice of a point of view is very important for understanding a scene.

In Fig. 3.1, one can see two different views of the same scene. Only the second one could help to understand it.

Obviously, it is not possible to try all the possible angles of view and to choose the best one because the camera's movement space is continuous. Even if we decide to sample the camera's space, we are typically in a case where there are many possibilities of solution and we need a method to choose a good enough solution. In such cases the use of artificial intelligence technique of heuristic search is recommended.

D. Plemenos and G. Miaoulis: Visual Complexity, SCI 200, pp. 65–94.

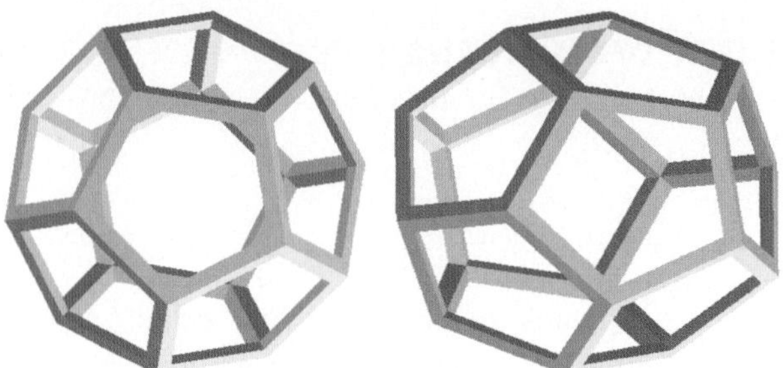

Fig. 3.1: Bad and good point of view for a scene

In such a case, it is again better to ask the computer work for us. If the view quality criterion for a point of view is based on the scene geometry, the computer (more precisely the program of rendering) knows better than the user the geometry of the scene and can evaluate several points of view much faster. So, the better way to well choose a good point of view is to test "all" possible points of view and to choose the better. Of course, it is impossible to test "all" possible points of view because there is an infinity of them and, for this reason, Artificial Intelligence techniques like heuristic search may be used in order to test only "pertinent" points of view.

3.2   Static scene understanding

One of the main goals of 3D scene rendering is to understand a scene from its rendered image. However, the rendered image of the scene depends on the choice of the point of view. If the point of view is a bad one, the scene may be very difficult to understand from its image. Interactively choosing a point of view for a scene is a very difficult and human time consuming process because on the one hand the user has not a good knowledge of the scene and, on the other hand, it is not easy for the user to select a point of view from a 3D scene from a 2D screen.

As the computer has a more complete knowledge of the scene, it is much better to ask the computer to find an interesting point of view. To do this, it is necessary to define a view quality criterion, in order to give the computer a way to evaluate points of view and to choose the better one.

As there is an infinity of potential points of view for a given scene, we also need a way to evaluate only a finite number of points of view and, even, to avoid evaluation of useless ones.

The first general method of automated computation of a good viewpoint for a polygonal scene was proposed in 1991 [2] and improved in 1996 [3]. This method will be described in a section 3.5.

3.3 Non degenerated views

Kamada et al. [4] consider a direction as a good point of view if it minimizes the number of degenerated images of objects when the scene is projected orthogonally. A degenerated image is an image where more than one edges belong to the same straight line. In Fig. 3.2, on can see a good view of a cube together with degenerated views of it.

The used method avoids the directions parallel to planes defined by pairs of edges of the scene.

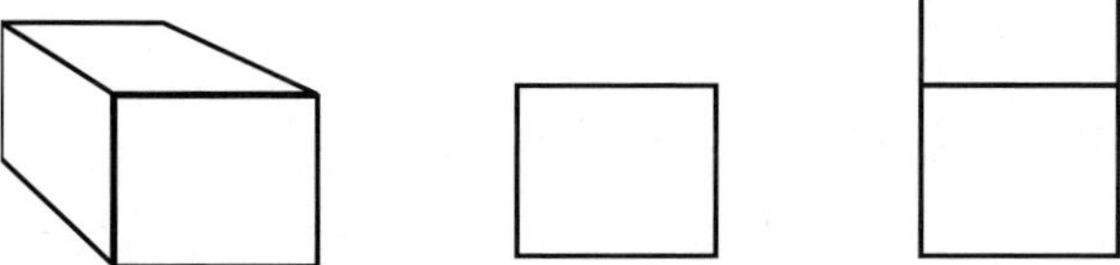

Fig. 3.2: good and degenerated views of a cube

If L is the set of all the edges of the scene and T the set of unit normal vectors to planes defined by couples of edges, let's call $\vec{P}$ the unit vector of the direction of view to be computed.

In order to minimize the number of degenerated images, the angles of the vector $\vec{P}$ with the faces of the scene must be as great as possible. This means that the angles of the vector $\vec{P}$ with the elements of T must be as small as possible. The used evaluation function is the following:

$$f(\vec{P}) = \text{Min}_{\vec{t} \in T}(|\vec{P} \cdot \vec{t}|)$$

As the purpose is to minimize the angle between $\vec{P}$ and $\dot{t}$, we must maximize the angle between these two vectors. So, we must maximize the function $f(\vec{P})$. To do this, a vector $\vec{P}$ which minimizes the greater angle between itself and the elements of T must be found.

In order to compute this vector, the authors have proposed two methods. The first one is simple and its main advantage is to be very fast. The second method is more precise because it is based on the computation of all the normal vectors to all the faces of a scene and all the angles between these vectors. This method is time consuming. Only the first method will be explained here.

As the purpose of this method is to decrease the computation cost of the function $f(\vec{P})$ for all the unit normal vectors, a set E of uniformly distributed unit vectors is chosen on the unitary sphere defined at the center of the scene.

The function $f(\vec{P})$ is computed for each element of the set E and the vector with the maximum value is chosen.

This method is interesting enough due to its rapidity. Its precision depends on the number of elements of the set E.

The technique proposed by Kamada is very interesting for a wire-frame display. However it is not very useful for a more realistic display. Indeed, this technique does not take into account the visibility of the elements of the considered scene and a big element of the scene may hide all the others in the final display.

3.4   Direct approximate viewpoint calculation

Colin [4] has proposed a method initially developed for scenes modeled by octrees. The purpose of the method was to compute a good point of view for an octree.

The method uses the principle of "direct approximate computation" to compute a good direction of view. This principle can be described as follows:

1.   Choose the three best directions of view among the 6 directions corresponding to the 3 coordinates axes passing through the center of the scene.
2.   Compute a good direction in the pyramid defined by the 3 chosen directions, taking into account the importance of each one of the chosen directions.

How to estimate the importance of a direction of view? What is a good criterion to estimate the quality of a view? The idea is to maximize the number of visible details. A direction of view is estimated better than another one if this direction of view allows to see more details than the other.

In the method of Colin, each element of the scene owns a "weight" with reference to a point of view. This weight is defined as the ratio between the visible part of the element and the whole element. It allows to attribute an evaluation mark to each direction of view and to use this mark to choose the three best directions.

When the three best directions of view d1, d2 and d3 have been chosen, if m1, m2 and m3 are the corresponding evaluation marks, a "good" direction of view can be computed by linear interpolation between the three chosen directions of view, taking into account the respective evaluation marks. So, if m1>m2>m3, the final "good" direction is computed (Fig. 3.3) by rotating the initial direction d1 to the direction d2 by an angle j1=90*m1/(m1+m2), then to the direction d2 by an angle j2=90*m1/(m1+m3).

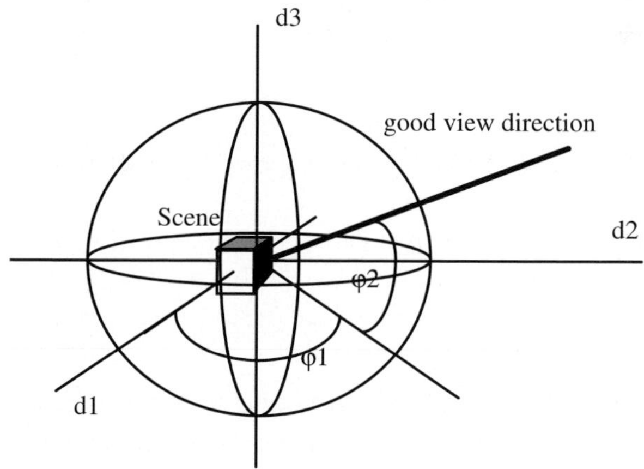

Fig. 3.3: direct approximate computation of a good direction of view

The used good view criterion for this method is only defined for octrees and it is a function of the angle of view of each voxel of the scene from the point of view.

3.5 Iterative viewpoint calculation

This good point of view computing method, proposed by Plemenos [2, 3], was developed and implemented in 1987 but it was first published only in 1991. Improvements of this method, integrated in this section, may be found in [6, 7, 8, 9].

The process used to determine a good point of view works as follows:

The points of view are supposed to be on the surface of a virtual sphere whose the scene is the center. The surface of the sphere of points of view is divided in 8 spherical triangles (Fig. 3.4).

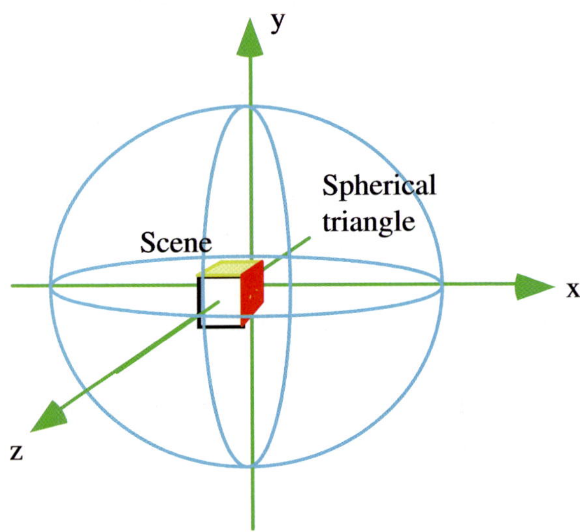

Fig. 3.4: sphere divided in 8 spherical triangles

The best spherical triangle is determined by positioning the camera at each intersection point of the three main axes with the sphere and computing its importance as a point of view. The three intersection points with the best evaluation are selected. These three points on the sphere determine a spherical triangle, selected as the best one.

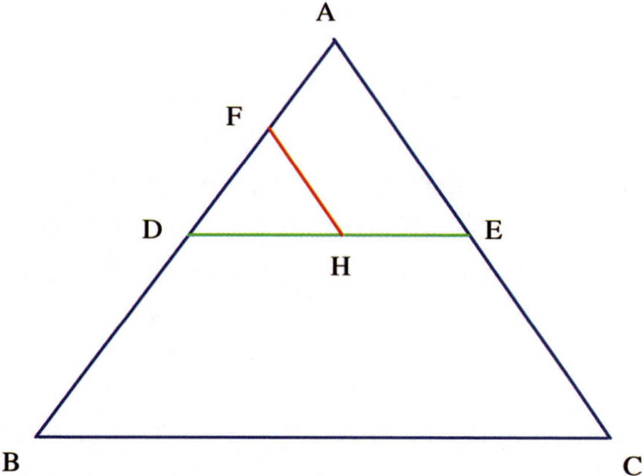

Fig. 3.5: Heuristic search of the best point of view by subdivision of a spherical triangle. From an initial spherical triangle ABC, a new spherical triangle ADE is computed and so on. The vertices of spherical triangles represent points of view.

The next problem to resolve is selection of the best point of view on the best spherical triangle. The following heuristic search technique is used to resolve this problem:

If the vertex A (Fig. 3.5) is the vertex with the best evaluation of the spherical triangle ABC, two new vertices E and F are chosen at the middles of the edges AB and AC respectively and the new spherical triangle ADE becomes the current spherical triangle. This process is recursively repeated until the quality of obtained points of view does not decrease. The vertex of the final spherical triangle with the best evaluation is chosen as the best point of view.

This heuristic search-based best viewpoint calculation gives always good results, even if, from the theoretical point of view, it is possible that it fails in some cases. Indeed, let's consider the first step of subdivision of a spherical triangle in Fig. 3.6. The area BCED will never be explored.

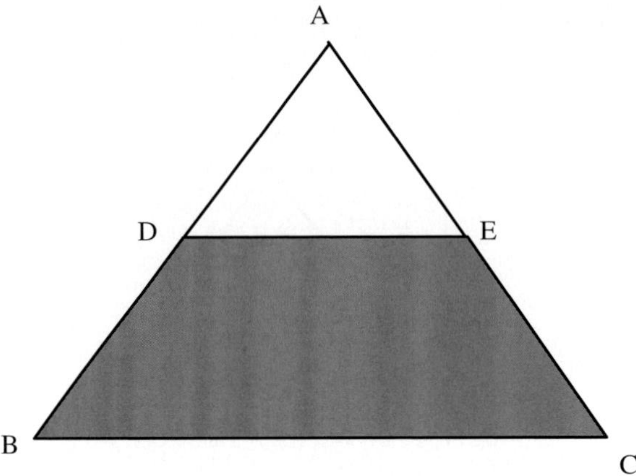

Fig. 3.6: First subdivision of the spherical triangle ABC. The area BCED is not explored

If we consider the second step of subdivision, that is the subdivision of the spherical triangle ADE, an additional area, AFHE in Fig. 3.7, will never be explored.

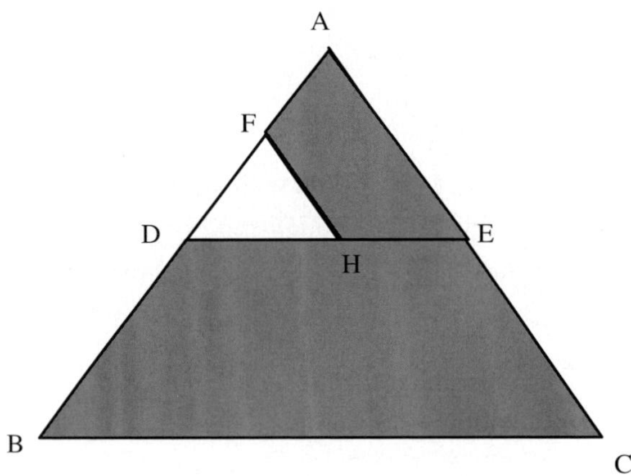

Fig. 3.7: Areas BCED and AFHE will never be explored

This possible drawback can be suppressed by subdividing the current spherical triangle in 4 new spherical triangles (Fig. 3.8) and so on. At the end of the process, the found viewpoints are compared among them and the best one is selected as the best viewpoint.

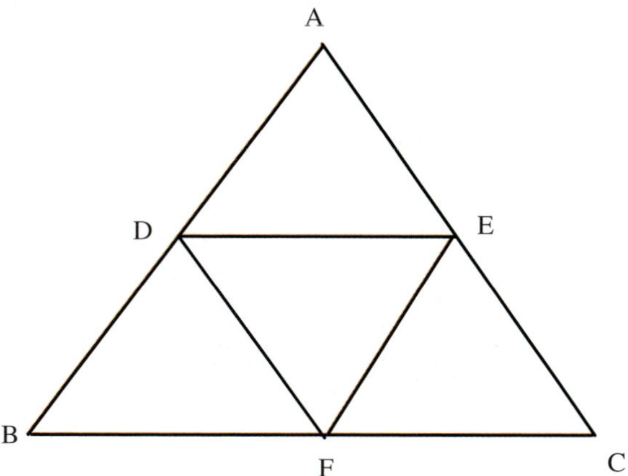

Fig. 3.8: The current spherical triangle is divided in 4 new spherical triangles

However, we think that this accuracy improvement is not necessary because the used method is accurate enough and very fast, whereas the improved method is time consuming.

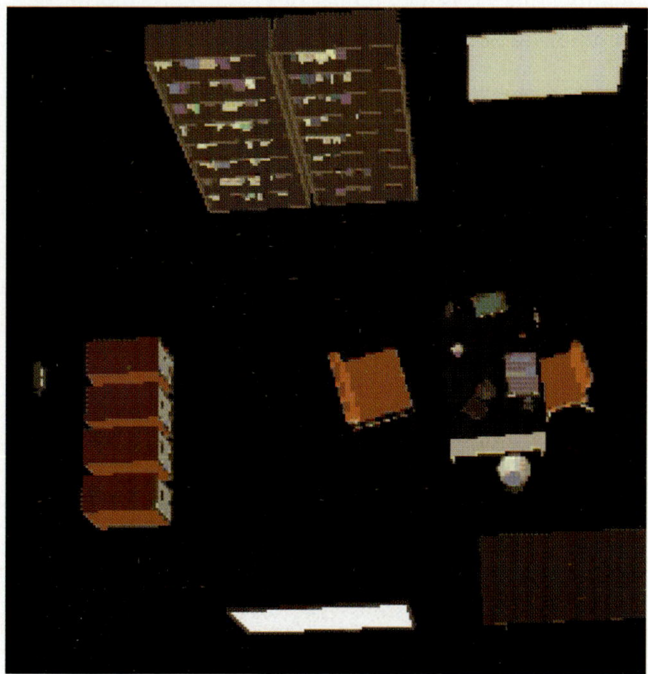

Fig. 3.9: Best viewpoint computed for a scene representing an office (Image G. Dorme)

Fig. 3.9 shows the best view of a scene representing an office, obtained by the method of iterative viewpoint calculation.

3.5.1 The viewpoint quality criterion

The initially proposed viewpoint quality criterion was the estimated number of visible polygons of the scene. The best viewpoint was the one permitting to see a maximum of polygons of the scene. Obtained results were satisfactory but, in some cases, the retained criterion does not allow the user to well understand the scene.

Thus, let's consider the scene of figure 3.10, composed of three objects. The view direction computed with the retained criterion gives the image of figure 3.11. This view shows a maximum of faces but it doesn't allow the user to have a good knowledge of the scene.

For this reason, a new viewpoint quality criterion was proposed. This new viewpoint quality criterion is the number of visible polygons of the scene from the given point of view, together with the total visible area of the scene projected on the screen.

Fig. 3.10: A scene composed of 3 objects

Fig. 3.11: Selected view for the scene of Fig. 3.10

All potential viewpoints are supposed to be on the surface of a sphere surrounding the scene (Fig. 3.4). Heuristic search is used to choose viewpoints only in potentially interesting regions, obtained by subdivision of spherical triangles, as it was described above (Fig. 3.5).

More precisely, the importance of a point of view will be computed using the following formula:

$$I(V) = \frac{\sum_{i=1}^{n}\left[\frac{P_i(V)}{P_i(V)+1}\right]}{n} + \frac{\sum_{i=1}^{n} P_i(V)}{r} \qquad (1)$$

where: I(V) is the importance of the view point V,
Pi(V) is the projected visible area of the polygon number i
obtained from the point of view V,
r is the total projected area,
n is the total number of polygons of the scene.

In this formula, [a] denotes the smallest integer, greater than or equal to a, for any a.

3.5.2 Fast viewpoint quality calculation

In practice, these measures are computed in a simple manner, with the aid of graphics hardware using OpenGL [6, 9]. A different color is assigned to every face, an image of the scene is computed using integrated z-buffer and a histogram of the image is computed. This histogram gives all information about the number of visible polygons and visible projected area of each polygon.

In order to compute information needed by this equation, OpenGL and its integrated z-buffer is used as follows:

A distinct colour is given to each surface of the scene and the display of the scene using OpenGL allows to obtain a histogram (Fig. 3.12) which gives information on the number of displayed colours and the ratio of the image space occupied by each colour.

As each surface has a distinct colour, the number of displayed colours is the number of visible surfaces of the scene from the current position of the camera. The ratio of the image space occupied by a colour is the area of the projection of the viewpoint part of the corresponding surface. The sum of these ratios is the projected area of the visible part of the scene. With this technique, the two viewpoint complexity criteria are computed directly by means of an integrated fast display method.

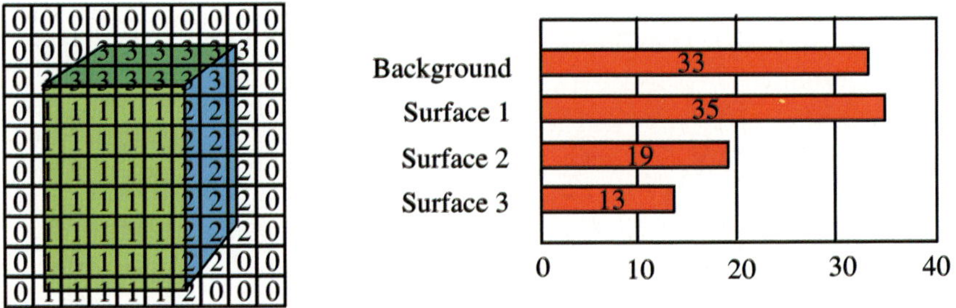

Fig. 3.12: Fast computation of number of visible surfaces and area of projected viewpoint part of the scene by image analysis

3.6 Information theory-based viewpoint calculation

Sbert et al. [6] proposed to use information theory in order to establish an accurate criterion for the quality of a point of view. A new measure is used to evaluate the amount of information captured from a given point of view. This measure is called *viewpoint entropy*. To define it, the authors use the relative area of the projected faces over the sphere of directions centered in the point of view (Fig. 3.13).

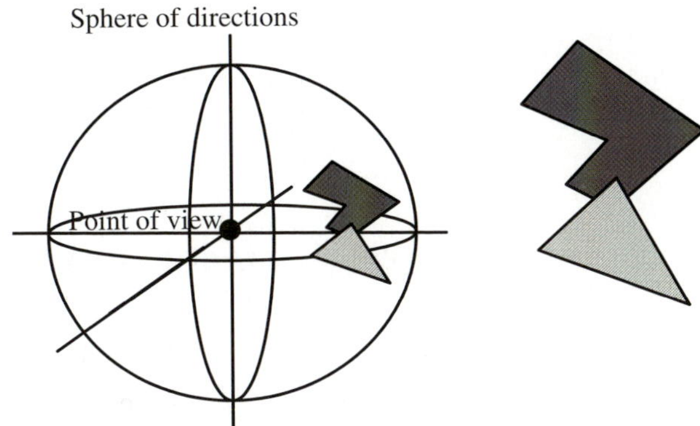

Fig. 3.13: the projected areas of polygons are used as probability distribution of the entropy function

The viewpoint entropy is then given by the formula:

$$H_p(X) = -\sum_{i=0}^{N_f} \frac{A_i}{A_t} \log\frac{A_i}{A_t} \qquad (2)$$

where Nf is the number of faces of the scene, Ai is the projected area of the face i and At is the total area covered over the sphere.

The maximum entropy is obtained when a viewpoint can see all the faces with the same relative projected area Ai/At. The best viewpoint is defined as the one that has the maximum entropy.

Fig. 3.14: point of view based on viewpoint entropy (Image P.-P. Vazquez)

To compute the viewpoint entropy, the authors use the technique proposed in [5] and described above, based on the use of graphics hardware using OpenGL.

The selection of the best view of a scene is computed by measuring the viewpoint entropy of a set of points placed over a sphere that bounds the scene. The point of maximum viewpoint entropy is chosen as the best one. Fig. 3.14 presents an example of results obtained with this method.

3.7   Total curvature-based viewpoint calculation

Most of the better known methods using the notion of viewpoint complexity to evaluate the pertinence of a view are based to two main geometric criteria: number of visible polygons and area of the projected visible part of the scene. Thus, equation (1) of section 3.5 is often used to evaluate the viewpoint complexity for a given scene.

However, even if the methods using these criteria give generally interesting results, the number of polygons criterion may produce some drawbacks. Indeed, let us consider a scene made from a single polygon (see Fig. 3.15). This polygon may be subdivided in several other polygons and, in such a case, the number of visible polygons will depend on the number of subdivisions of the initial polygon. A viewpoint complexity evaluation function will give different results for the same scene, according to its subdivision degree.

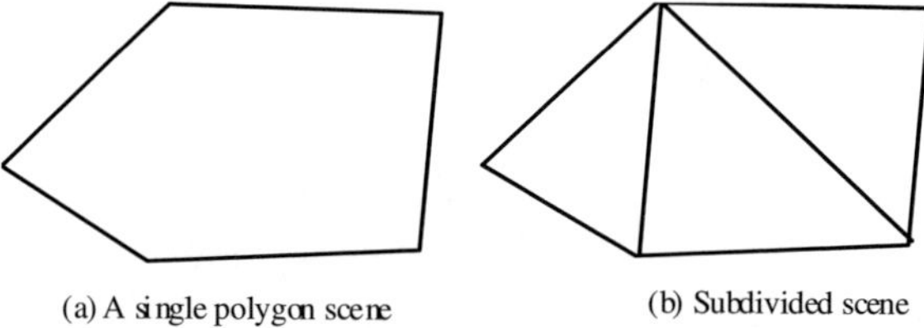

Fig. 3.15: The view quality for the same scene and from the same viewpoint depends on the subdivision level of the scene.

In order to avoid this drawback, another criterion was proposed by Sokolov et al. [11, 12, 13], which takes into account the curvature of the scene. More precisely, the number of polygon criterion is replaced by the criterion of total curvature of the scene. In the proposed method, the importance of a view from a viewpoint p is given by the equation:

$$I(p) = \sum_{v \in V(p)} \left| 2\pi - \sum_{\alpha_i \in \alpha(v)} \alpha_i \right| \sum_{f \in F(p)} P(f) \tag{3}$$

where:
F(p) is the set of polygons visible from the viewpoint p,
P(f) is the projected area of polygon f,
V(p) is the set of visible vertices of the scene from p,
$\alpha(v)$ is the set of angles adjacent to the vertex v.

Equation 3 uses the curvature in a vertex (Fig. 3.16), that is the sum of angles adjacent to the vertex minus 2π.

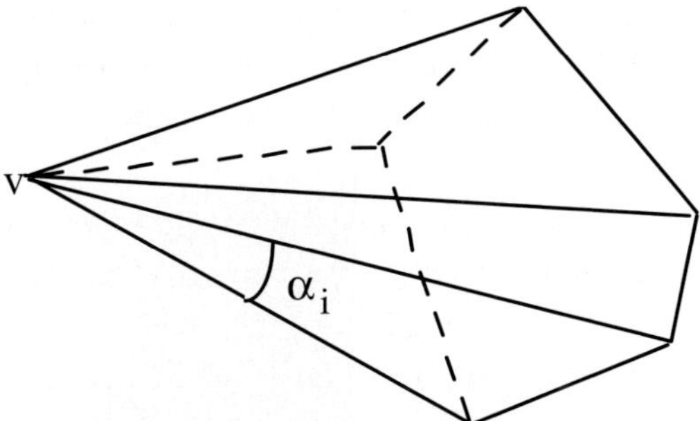

Fig. 3.16: Curvature in a vertex

The main advantage of the proposed criterion is that it is invariant to any subdivision of the scene elements maintaining the topology. Another advantage is that it can be extended in order to use the total integral curvature of curved surfaces.

The best viewpoint is computed by using a data structure, so-called *visibility graph*, witch allows to associate to every discrete potential viewpoint on the surface of the surrounding sphere, the visual pertinence of the view from this viewpoint. Fig. 3.17 shows the best point of view for a sphere with holes, containing various objects.

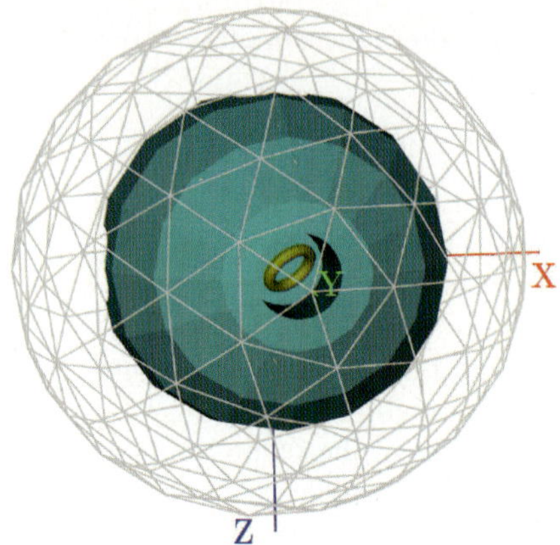

Fig. 3.17: Best viewpoint computation for the scene, using the total curvature criterion instead of the number of polygons one (Image Dmitry Sokolov).

Fig. 3.18 shows the best viewpoint for a scene test scene representing an office. The best viewpoint is calculated using the total curvature criterion.

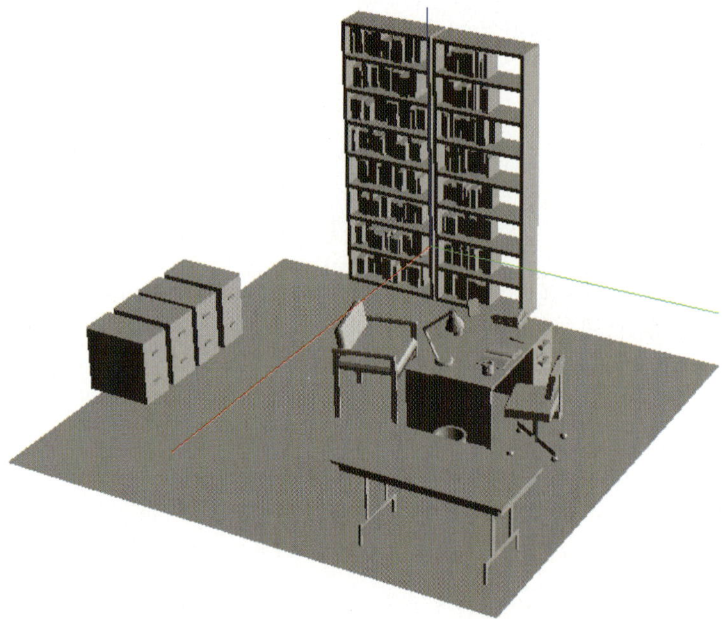

Fig. 3.18: The best viewpoint for the office scene, based on the total curvature criterion (Image Dmitry Sokolov).

In Fig. 3.19 one can see a comparison between a viewpoint computed with the method presented in section 3.5 (on the left) and a viewpoint computed with the total scene curvature method (on the right).

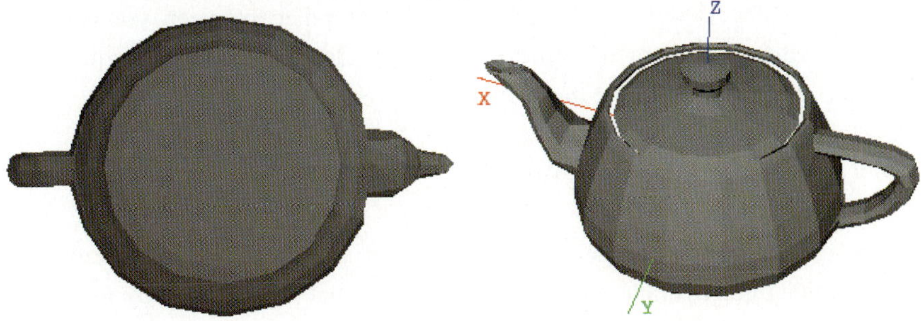

Fig. 3.19: Maximum projected area (left) vs total scene curvature (right)
(Images Dmitry Sokolov)

Fig. 3.20 presents a comparison between a viewpoint computed with the method presented in section 3.6 (on the left) and a viewpoint computed with the total scene curvature method (on the right).

Fig. 3.20: Viewpoint entropy (left) vs total scene curvature (right)
(Images Dmitry Sokolov)

3.8   Mesh Saliency-based viewpoint calculation

The concept of mesh saliency was introduced in 2005 by Chang Ha Lee et al. [14]. The goal of this concept is to bring perception-based metrics in evaluation of the pertinence of a view. According to the authors, a high-curvature spike in the middle of a largely flat region is perceived to be as important as a flat region in the middle of densely repeated high-curvature bumps.

Mesh saliency is defined as the absolute difference between the Gaussian-weighted averages computed in fine and coarse scales. The following equation resumes the method to compute mesh saliency $S(v)$ of a vertex v.

$$S(v) = |G(C(v),\sigma) - G(C(v),2\sigma)|$$

In this equation $C(v)$ is the mean curvature of vertex v. $G(C(v),\sigma)$ denotes the Gaussian-weighted average of the mean curvature and is computed by the following equation:

$$G(C(v),\sigma) = \frac{\sum_{x \in N(v,2\sigma)} C(x)\exp[-\|x-v\|^2/(2\sigma^2)]}{\sum_{x \in N(v,2\sigma)} \exp[-\|x-v\|^2/(2\sigma^2)]}$$

$N(v,\sigma)$, in this equation is the neiborhood for a vertex v and may be defined as:

$$N(v,\sigma) = \{x| \ \|x-v\|<\sigma, x \text{ is a mesh point}\}$$

The mesh saliency can be computed at multiple scales. So, the saliency of a vertex v at a scale level i is computed by the formula:

$$S_i(v) = |G(C(v),\sigma_i) - G(C(v),2\sigma_i)|$$

where σ_i is the standard deviation of the Gaussian filter at scale i.

The authors use mesh saliency to compute interesting points of view for a scene. The shown examples seem interesting. An important advantage of the method is that the notion of mesh saliency is defined and may be computed at multiple scales.

3.9 Object understanding-based viewpoint calculation

In some cases, it is possible to add semantics to a scene and to decompose it into a set of objects, corresponding to well known features of each part of the scene.

Fig. 3.21 shows an example of such a scene, where the decomposition into a set of objects is shown by different colors. The scene is made of many different objects: the computer case, the display, the mouse, the mouse pad, two cables, the keyboard and several groups of keys.

In the scene of Fig. 3.21, only 20% of the display surface is visible, but it is not a problem for a human to recognize it, because if we are familiar with the object, we could *predict* what does it look like. Thus, we could conclude that if there exists a proper decomposition of a scene into a set of objects, the visibility of the objects could bring more information than just the visibility of the faces [11, 12, 13].

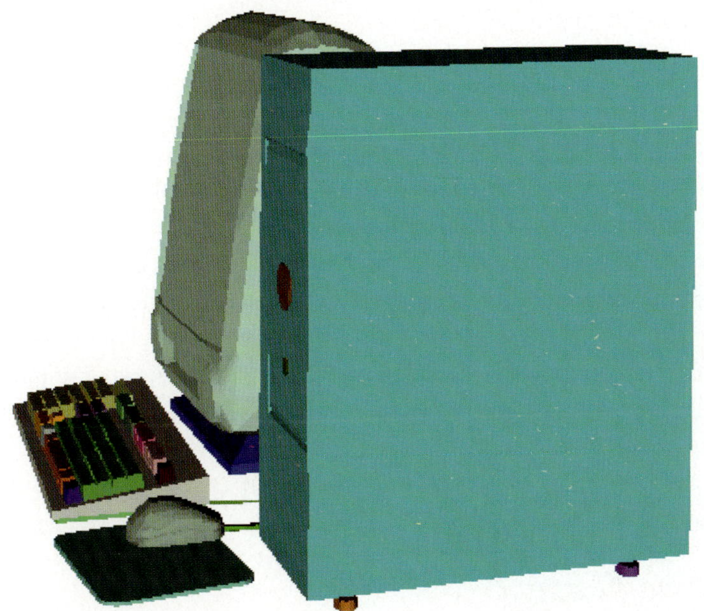

Fig. 3.21: A scene which may be decomposed in objects (Image by Dmitry Sokolov)

Of course, it is not always easy to decompose any scene in objects but there are many scenes which consist of non adjacent primitives and their decomposition could be performed automatically. An area where scene decomposition is quite natural is Declarative Modeling by Hierarchical Decomposition (DMHD) [2, 15]. The method presented in this section could be easily used with scenes by such a declarative modeler.

Let us suppose that for each object ω of a scene, its importance is known. For example, if the method is incorporated into some dedicated declarative modeler, the importance of an object could be assigned in dependence on its functionality.

If no additional information is provided and the user takes into account scene geometry only, then the size of the object bounding box could be considered as the importance function:

$$q(\omega) = \max_{u,v \in V_\omega} |u_x - v_x| + \max_{u,v \in V_\omega} |u_y - v_y| + \max_{u,v \in V_\omega} |u_z - v_z|$$

where V_ω is the set of vertices of the object ω.

It is also necessary to introduce a parameter characterizing the *predictability* (or *familiarity*) of an object: ρ_ω. This parameter determines the quantity of object surface to be observed in order to well understand what the object looks like. If an object is well-predictable, then the user can recognize it even if he sees a small part of it. Bad predictability forces the user to observe attentively all the surface.

Finally, the proposed evaluation function of the viewpoint quality is the weighted sum of scene objects importances, with respect to their visibility:

$$I(p) = \sum_{\omega \in \Omega} q(\omega) \cdot \frac{\rho_\omega + 1}{\rho_\omega + \theta_{p,\omega}} \theta_{p,\omega}$$

where $\theta_{p,\omega}$ is the fraction of visible area for the object ω from the viewpoint p. $\theta_{p,\omega} = 0$ if the object is not visible and $\theta_{p,\omega} = 1$ if one can see all its surface from the viewpoint p.

The quantity:

$$\frac{\rho_\omega + 1}{\rho_\omega + \theta_{p,\omega}} \theta_{p,\omega}$$

represents the visual understanding quality of an object.

In Fig. 3.22 one can see the best viewpoint for the computer scene, obtained by using the quality criterion described in this section. This viewpoint is very close to the viewpoint obtained with the quality criterion of section 3.7. The main difference is that other viewpoints, as for example the one of Fig. 3.21, considered bad with the criterion of section 3.7 are now considered good enough, because they allow to well understand the scene.

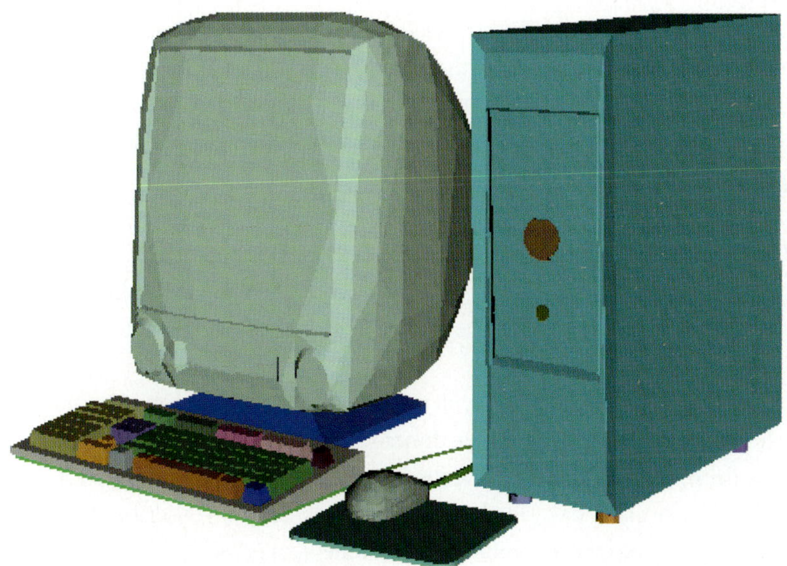

Fig. 3.22: Best viewpoint obtained with the object understanding criterion
(Image by Dmitry Sokolov)

3.10 A priori viewpoint calculation

All the visual scene understanding methods presented above try to compute a good point of view (if possible the best one), in order to allow a human user to well understand a given scene. All these methods suppose that the geometry of the scene to be processed is entirely known. We call them "a posteriori" methods.

However, in some cases, especially during the scene designing process, the user doesn't know exactly the geometry of the scene to design. This is because he (she) has still a vague idea of the scene to be obtained and so he (she) expects the modeler to help him to give a precise form to this idea. This is especially true for declarative modeling [1, 2, 15, 18]. In this scene modeling technique, the user can use general and often imprecise properties to describe his (her) scene.

The Declarative Modeling by Hierarchical Decomposition (DMHD) paradigm of declarative scene modeling [2, 15] will be briefly presented below, in order to explain how it is possible to introduce scene understanding techniques, based on general properties of a scene, without any knowledge of the scene geometry. We will call these scene understanding techniques "a priori" methods. The interested reader may find more details on the DMHD declarative modeling paradigm in chapter 2 of this volume.

The Declarative Modeling by Hierarchical Decomposition (DMHD) technique allows the user to design a scene in top-down manner, by using high level properties in various levels of detail. The principle of DMHD is the following:

If a scene is easy to describe, it is described by a small number of properties which can actually be **size** (inter-dimensions) **properties** (higher than large, as high as deep, etc.) or **form properties** (elongated, very rounded, etc.). Otherwise the scene is partially described with properties easy to describe and is then decomposed into a number of sub-scenes and the same description process is applied to each sub-scene. In order to express relationships within sub-scenes of a scene, **placement properties** (put on, pasted on the left, etc.) and **size** (inter-scenes) **properties** (higher than, etc.) are used.

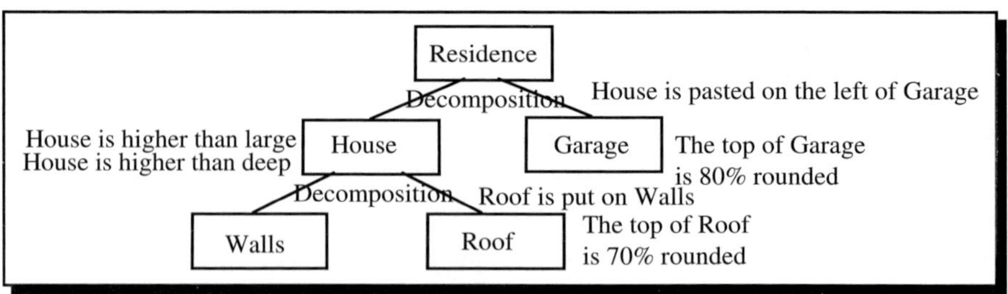

Fig. 3.23: Hierarchical declarative description of a scene

In Fig. 3.23, the scene named *Residence* is difficult to describe and is decomposed into two sub-scenes, *House* and *Garage*. House is also difficult to describe and is decomposed into *Walls* and *Roof*.

This user description is translated into an internal model, made up of a set of Prolog-like rules and facts.

Thus, the above top-down description will produce the following internal model :

```
Residence (x) ->        CreateFact(x)
                        House(y)
                        Garage(z)
                        PastedLeft(z,y);


House (x)       ->      CreateFact(x)
                        HigherThanLarge(x)
                        HigherThanDeep(x)
                        Walls(y)
                        Roof(z)
                        PutOn(z,y);


Garage (x) ->           CreateFact(x)
                        TopRounded(x,80);


Walls (x) ->   CreateFact(x);


Roof (x) ->    CreateFact(x)
               TopRounded(x,70);
```

In this set of rules appear properties like *TopRounded*, *PutOn*, etc. which are already defined into the modeler. Additional properties can be added to the knowledge base of the modeler by the expert user. The predicate *CreateFact* is automatically generated by the modeler and is used to manage hierarchical dependencies within elements of the scene.

A special inference engine, or a constraint resolution engine, processes the rules and facts of the internal model and generates all the solutions.

In such cases it is very important for the user to have a display showing the scene properties. As the modeler knows the general properties of the scene, it can compute this display by combining these properties. In this section, we will present techniques to automatically compute a good view direction by *a priori* techniques [2, 3].

Starting from the bounding box of an object (or a scene), we consider eight view directions, each of them passing from one vertex of the embedding box of the object and from the center of this embedding box (Fig. 3.24).

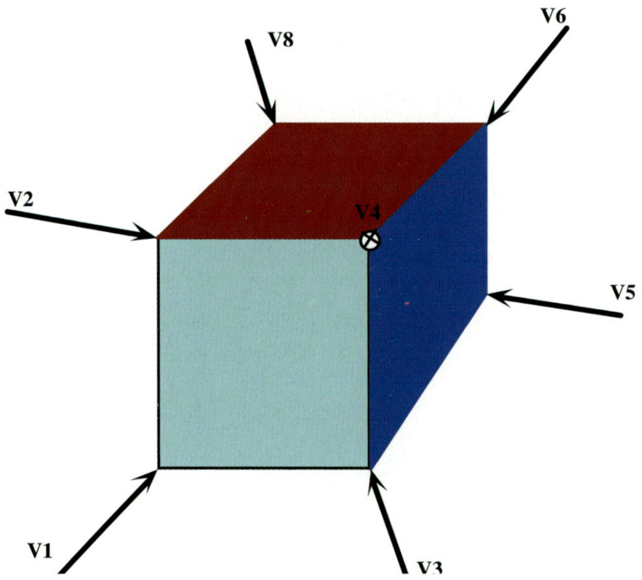

Fig. 3.24: Main view directions associated to an object

3.10.1 Processing of form properties

Each dissymetric form property invokes a number of these directions. For example, the property **Top Rounded** invokes the directions V2, V4, V6 and V8.

Unlike dissymetric form properties, symetric properties do not invoke any particular direction : any one of the 8 main directions is a good one.

During the generation process, the value of a counter of shown properties associated with each main view direction is modified according to the processed form property.

3.10.2 Processing of placement properties

Relative placement properties between two sub-scenes permit to compute a new good view direction for the scene which is the union of these sub-scenes. To do this, the following method is used (Fig. 3.25):

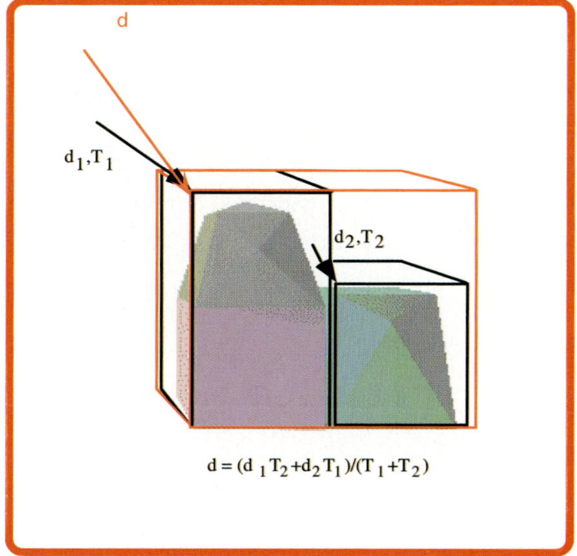

Fig 3.25: Good view direction for a union of objects

1.   Compute a new view direction **d** from the best ones, **d1** and **d2**, of the two sub-scenes. A view is better than another one if its counter of shown properties is greater than the counter of the other view direction.

     To determine the new view direction **d** we take into account the sizes **T1** and **T2** of the two sub-scenes, by using the formula :

$$d = \frac{T_1 d_2 + T_2 d_1}{T_1 + T_2}$$

If **n1** and **n2** are respectively the numbers of shown properties associated with **d1** and **d2**, the counter of shown properties associated with **d** will be **n**, where **n** is given by the formula:

$$n = \frac{T_1\, n_2 + T_2\, n_1}{T_1 + T_2}$$

2. Consider the bounding box of the whole scene and its 8 main directions (figure 6). Let us call αi the angle between **d** and the main direction **Vi** and α**min** the minimum of these angles:

$$\alpha\text{min} = (\min\,(\alpha i),\ i = 1, 2, \dots, 8).$$

If α**min** is equal to the angle between **d** and one main direction **Vj**, that is, to the angle α**j**, the main direction **Vj** is selected as the new view direction.

The number **nk** of properties associated to each main view direction will be inversely proportional to the angle between **d** and this direction:

$$n_k = \frac{n\left[\left(\sum\limits_{i=1}^{8} \alpha_i\right) - \alpha_k\right]}{\sum\limits_{i=1}^{8} \alpha_i}$$

where **n** is the number of properties associated with the direction **d**.

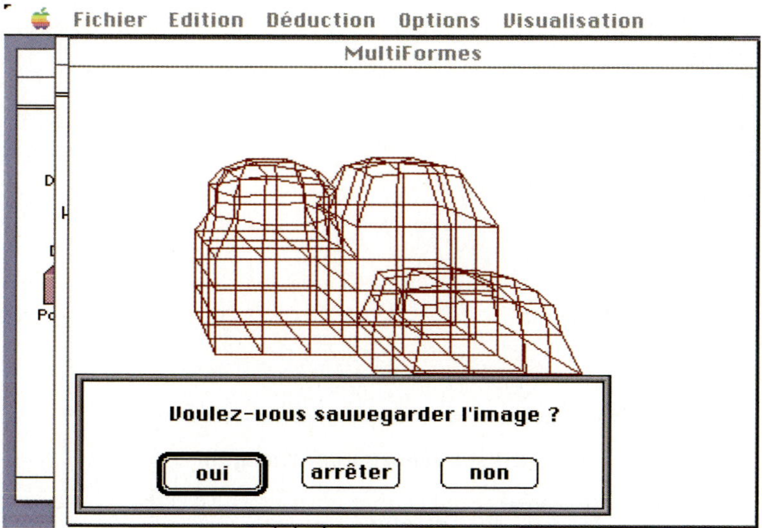

Fig. 3.26: Declaratively designed scene, seen from an arbitrary viewpoint

In Fig. 3.26 one can see a scene designed with the DMHD declarative scene modeling technique and displayed from an arbitrary point of view. Fig. 3.27 shows the same scene displayed from a point of view automatically computed by the above presented a priori viewpoint calculation technique.

This a priori method gives good results for wireframe display and allows scene understanding during the early phase of scene declarative design. However, like the Kamada a posteriori scene understanding method, it doesn't take into account visibility problems and, for this reason, it is not very useful for a realistic scene display.

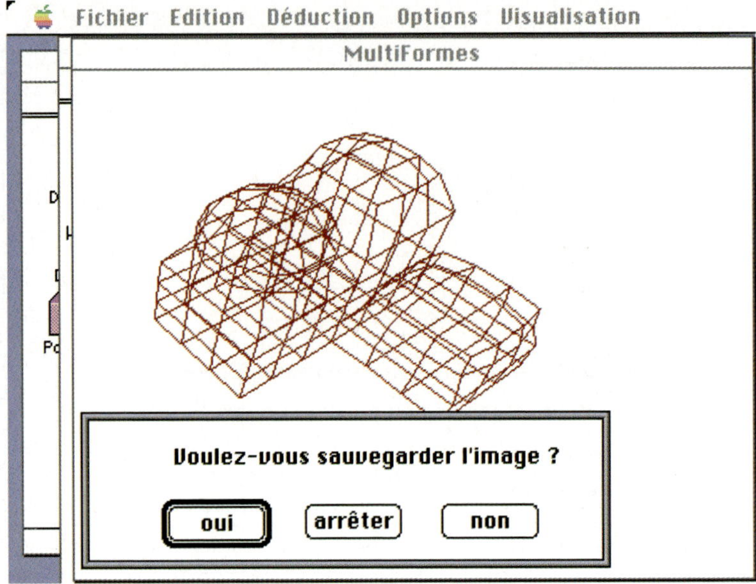

Fig. 3.27: The scene of Fig. 3.26, seen from an automatically computed viewpoint
by the a priori tecghnique presented in this section

3.11   Conclusion

In this chapter we have tried to present the problem of scene understanding and its
importance in Computer Graphics. This importance was not always well understood, especially before the end of years 1990.

With the development of Internet during the last decade, people started to better understand the importance of scene understanding and several papers were
published, especially after the beginning of the 21st century.

Scene understanding techniques presented in this chapter are based on automatic search of a good point of view, allowing a good understanding of a scene for
a human user. In order to achieve this goal, Artificial Intelligence techniques are
often used, especially heuristic search.

All the presented scene understanding techniques, except one, are based on
full knowledge of the scene geometry. We call these techniques *a posteriori*

techniques. They allow a good understanding of a scene from its realistic rendering using the computed point of view.

The last presented technique is an *a priori* technique, because it works without sufficient knowledge of the scene geometry and is based on the use of general, often imprecise form and placement properties, known during a declarative scene modeling process. Only a priori techniques may be used during the declarative scene modeling process, when geometry is not yet precisely defined. The main drawback of a priori techniques is that they don't insure that the computed point of view is the best one, because they don't take into account the scene geometry. So, it is possible that a big object hides many details of the scene in a display from the computed point of view.

Some other methods, close to the ones presented in this chapter, were proposed during this last decade. Some recent methods try to take into account not only the geometry but also the lighting of the scene in order to evaluate the quality of a point of view [10, 14]. We are going to present such methods in chapter 5.

Of course, if visual static understanding from an only computed point of view is enough for simple scenes, it is not sufficient for complex ones. In such scenes, or virtual worlds, intelligent virtual world exploration strategies have to be defined in order to offer to a human observer a real possibility to understand them. Such virtual world exploration techniques will be presented in chapter 4.

References

1. Lucas, M., Martin, D., Martin, P., Plemenos, D.: The ExploFormes project: some steps towards declarative modeling of forms. In: AFCET-GROPLAN Conference, Strasbourg, France, November 29 - December 1, vol. 67, pp. 35–49 (1989) (in French); published in BIGRE
2. Plemenos, D.: A contribution to study and development of scene modeling, generation and display techniques - The MultiFormes project. Professorial Dissertation, Nantes, France (November 1991) (in French)
3. Plemenos, D., Benayada, M.: Intelligent Display in Scene Modeling. New Techniques to Automatically Compute Good Views. In: International Conference GraphiCon 1996, St Petersbourg, Russia, July 1 - 5 (1996)
4. Kamada, T., Kawai, S.: A Simple Method for Computing General Position in Displaying Three-dimensional Objects. Computer Vision, Graphics and Image Processing 41 (1988)
5. Colin, C.: Automatic computing of good views of a scene. In: MICAD 1990, Paris, France (February 1990)

6. Barral, P., Dorme, G., Plemenos, D.: Visual understanding of a scene by automatic movement of a camera. In: GraphiCon 1999, Moscow, Russia, August 26 - September 1 (1999)
7. Barral, P., Dorme, G., Plemenos, D.: Visual understanding of a scene by automatic movement of a camera. In: Eurographics 2000 (2000)
8. Barral, P., Dorme, G., Plemenos, D.: Visual understanding of a scene by automatic movement of a camera. In: International conference 3IA 2000, Limoges, France, March 3-4 (2000)
9. Dorme, G.: Study and implementation of understanding techniques for 3D scenes, Ph. D. thesis, Limoges, France, June 21 (2001)
10. Vázquez, P.P., Feixas, M., Sbert, M., Heidrich, W.: Viewpoint Selection Using Viewpoint Entropy. In: Vision, Modeling, and Visualization 2001, Stuttgart, Germany, pp. 273–280 (2001)
11. Sokolov, D., Plemenos, D., Tamine, K.: Methods and data structures for virtual world exploration. The Visual Computer 22(7), 506–516 (2006)
12. Sokolov, D.: Online exploration of virtual worlds on the Internet, Ph. D thesis, Limoges, France, December 12 (2006)
13. Sokolov, D., Plemenos, D.: Virtual world exploration by using topological and semantic knowledge. The Visual Computer 24(3) (March 2008)
14. Lee, C.H., Varshney, A., Jacobs, D.W.: Mesh saliency. ACM TOG 24(3), 659–666 (2005)
15. Plemenos, D.: Declarative modeling by hierarchical decomposition. The actual state of the MultiFormes project. In: International Conference GraphiCon 1995, St Petersbourg, Russia, July 3-7 (1995)
16. Lam, C., Plemenos, D.: Intelligent scene understanding using geometry and lighting. In: 10th International Conference 3IA 2007, Athens, Greece, May 30-31, pp. 97–108 (2007)
17. Vázquez, P.P., Sbert, M.: Perception-based illumination information measurement and light source placement. In: Proc. of ICCSA 2003. LNCS (2003)
18. Lucas, M., Desmontils, E.: Declarative modelers. Revue Internationale de CFAO et d'Infographie 10(6), 559–585 (1995) (in French)

4 Virtual World Exploration

Abstract. Automatic virtual world exploration techniques, allowing the user to understand complex unknown virtual worlds, will be presented in this chapter. Exploration is made by a virtual camera and is based on visual as well as scene covering criteria. The used techniques may be *global*, if the camera remains outside the explored virtual world, or *local*, if the camera moves inside the virtual world. Online and offline exploration methods are also presented.

Keywords. Good point of view, Viewpoint complexity, Viewpoint entropy, Scene geometry, Visibility, Total scene curvature, Virtual world, Virtual camera, Global exploration, Local exploration.

4.1 Introduction

In Chapter 3, we have seen why scene understanding is a very important area in Computer Graphics. Then, we have presented several techniques which may help the user to understand an unknown, or a badly known, scene, by automatically computing a good point of view for this scene.

Generally, computation of a single point of view is enough to understand a simple scene. However, for complex scenes, a single point of view doesn't give sufficient knowledge to understand them. In case of complex scenes (we call them *virtual worlds*) more than one point of view are necessary to well understand them. In reality, even a lot of points of view cannot insure a good understanding of a scene, because it doesn't give a global view of it. So, other methods have to be invented to carry out virtual world understanding.

Virtual worlds exploration techniques become nowadays more and more important. When, more than 10 years ago, we have proposed the very first methods permitting to improve the knowledge of a virtual world [2, 3], many people thought that it was not an important problem. People begun to understand the importance of this problem and the necessity to have fast and accurate techniques for a good exploration and understanding of various virtual worlds, only during these last years. However, there are very few papers facing this problem from the

D. Plemenos and G. Miaoulis: Visual Complexity, SCI 200, pp. 95–153.

computer graphics point of view, whereas several papers have been published on the robotics artificial vision problem.

The purpose of a virtual world exploration in computer graphics is completely different from the purpose of techniques used in robotics. In computer graphics, the goal of the program which guides a virtual camera is to allow a human being, the user, to understand a new world by using an automatically computed path, depending on the nature of the world. The main interaction is between the camera and the user, a virtual and a human agent and not between two virtual agents or a virtual agent and his environment.

It is very important to well understand the purpose of virtual world exploration presented in this chapter. Exploring virtual world, seen as a computer graphics problem, is quite different from the robotics problem where an autonomous agent moves inside a virtual or real world whose it is a part. In computer graphics, the main purpose of a virtual world exploration is to allow the user to understand this world. Here, the image of the world seen from the current point of view is very important because it will permit the user to understand the world. In robotics, the problem is different and the use of an image is not essential because the agent is not a human being.

There are two kinds of virtual world exploration. The first one is global exploration, where the camera remains outside the world to be explored. The second kind of exploration is local exploration. In such an exploration, the camera moves inside the scene and becomes a part of the scene. Local exploration can be useful, and even necessary, in some cases but only global exploration can give the user a general knowledge on the scene. In this chapter we are mainly concerned by global virtual world exploration. However, some interesting methods of local virtual world exploration will be presented, more precisely those, whose purpose is scene understanding by a human user. In any case, the purpose of this chapter is visual exploration of fixed unchanging virtual worlds.

On the other hand, in global as well in local virtual world exploration, two different exploration modes can be considered:

1. Real-time online exploration, where the virtual world is visited fofirst time and the path of the camera is computed in real-time, in an incremental manner. In this mode, it is important to apply fast exploration techniques, in order to allow the user to understand the explored world in real-time.

2. Offline exploration, where the camera path is computed offline and an exploration can be undertaken later. In a preliminary step, the virtual world is found and analyzed by the program guiding the camera movement, in order to determine interesting points of view and paths linking these points. The camera will explore the virtual world later, when necessary, following the determined path. In this exploration mode, it is less important to use fast techniques.

The chapter is organized in the following manner:

In sections 4.2, 4.3, 4.4 and 4.5 incremental online virtual world exploration techniques will be presented. More precisely, in section 4.2 a global online exploration technique will be presented, whereas section 4.3 will be dedicated to information theory-based global virtual world exploration. In section 4.4, a goal-based global online exploration technique will be described. The main principles of a local online virtual world exploration method will be explained in section 4.5.

In sections 4.6, 4.7 and 4.8 incremental offline virtual world exploration techniques will be presented. More precisely, in section 4.6, two global exploration techniques, based on minimal viewpoint set construction, will be presented, whereas in section 4.7 a total curvature based global exploration method will be described. A local exploration technique, based on object visibility, will be presented in section 4.8. Finally, we will conclude is section 4.9.

4.2 Global incremental online exploration

The purpose of the method presented below is to choose a trajectory for a virtual camera, allowing the user to have a good knowledge of an explored virtual world at the end of a minimal movement of the camera around it. The method is based on techniques for computing a good point of view, used together with various heuristics, in order to have smooth movement of the camera and to avoid fast returns to the starting point, due to local maximums. So, starting from a good viewpoint, the virtual camera moves on the surface of a sphere surrounding the scene, combining good view, smooth camera movement and distance from the starting point based heuristics [1, 2, 3, 4].

It must be clear that the notion of *good view direction* is a subjective concept. So an algorithm for computing such directions do not give a general solution but a solution good enough in many cases.

Unfortunately, the use of a single view direction is often insufficient to well understand a scene, especially in the case of a complex scene, that is a virtual world. Thus, in order to give the user sufficient knowledge of a virtual world, it should be better to compute more than one points of view, revealing the main properties of this virtual world. The problem is that, even if the solution of computing more than one view directions is satisfactory from a theoretical point of view, from the user's point of view it is not a solution at all because blunt changes of view direction are rather confusing for him (her) and the expected result (better knowledge of the explored virtual word) is not reached.

A way to avoid brutal changes of view direction is to simulate a virtual camera moving smoothly around the virtual world. This approach generates other problems because the determination of good views must be fast enough if we want to give the user an impression of online continuous movement of the camera. Moreover, sudden changes of the camera trajectory should be avoided in order to have a smooth movement of the camera, and heuristics should be provided to avoid attraction forces in the neighboring of a good view direction.

Some papers were published on the theme of camera movement but most of them study the problem of offline local camera movement. So, L. Fournier [7] gives a theoretical study of automatic smooth movement of a camera around a scene, based on reinforcement learning. J.-P. Mounier [8] uses genetic algorithms for planing paths for a camera walking inside a scene. In [9] a solver using interval arithmetic is used to obtain camera movements satisfying user-defined constraints.

As we have seen above, there are two techniques for understanding a virtual world by means of a virtual camera. In the first one the camera can **visit** the interior of the scene in order to have a closer view of the elements composing the scene. In the second technique the camera and the scene have two distinct nonintersecting universes. The camera **flies over** the scene choosing its trajectory in order to got a good knowledge of the scene properties with a minimum of displacement. In the following sections, a heuristic method to guide the movement of a virtual camera, based on the second technique, so-called global exploration, is presented.

4.2.1 A heuristic method to guide the movement of a camera

The camera moves on the surface of a sphere surrounding the scene (Fig. 4.1). The starting point of the camera is computed using the method of calculation of a good view direction described in chapter 3. In order to get faster calculation of a good view direction, an improvement based on the hardware capabilities is applied. This improvement was presented in chapter 3 and will be briefly described in sub-section 4.2.2. In future implementations we think that combination of the used good view criterion with the Kamada's one [2] would improve the quality of the computed view direction.

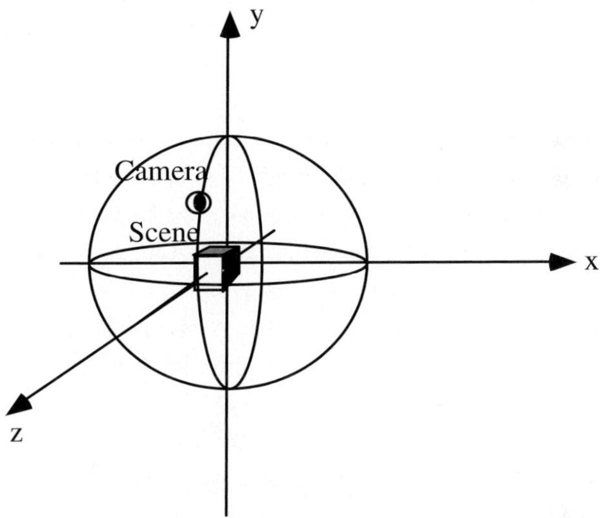

Fig. 4.1: The camera moves on the surface of a sphere

Once a good view direction has been computed and the corresponding view point has been determined on the surface of the sphere, the camera is set in this point and the movement of the camera starts by a research of the camera's initial movement direction. Initially, all directions are plausible and 8 displacement directions are considered (Fig. 4.2).

For each possible new position of the camera on the surface of the scene, corresponding to a displacement direction, the view direction from this point is evaluated and the chosen position is the one corresponding to the best view direction value.

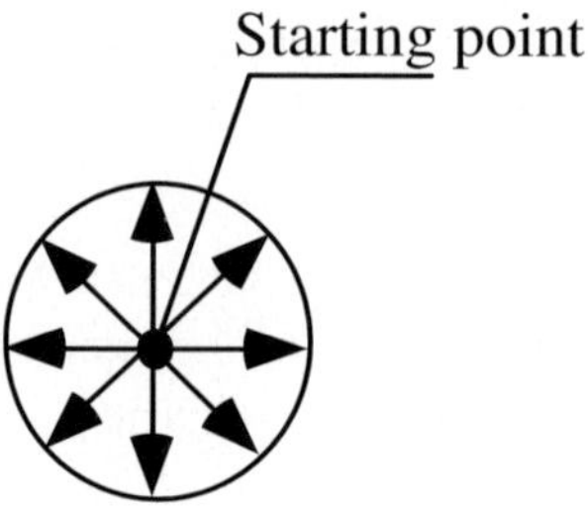

Fig. 4.2: Starting point and possible camera directions

After the first displacement of the camera, a movement direction is defined by the previous and the current position of the camera. As blunt changes of movement direction have to be avoided, in order to obtain a smooth movement of the camera, the number of possible new directions of the camera is reduced and only 3 directions are possible for each new displacement of the camera (Fig. 4.3).

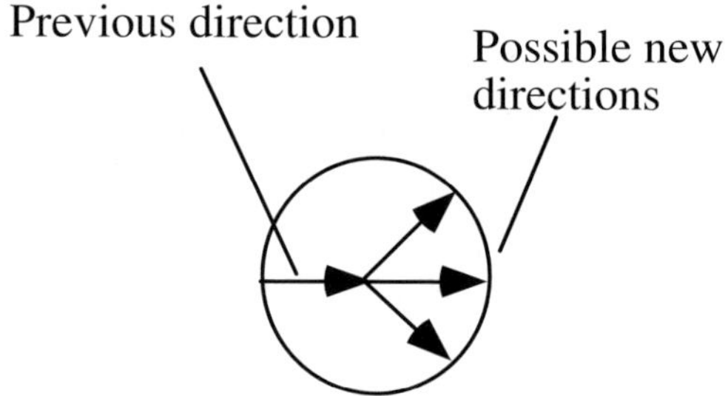

Fig. 4.3: Only 3 directions are considered for a smooth movement of the camera

One of the three possible directions is chosen using heuristic rules taking into account not only the view direction value of a point but also other parameters permitting to avoid attraction points and cycles in the camera's movement. These heuristic rules will be described in subsection 4.2.3.

In the current implementation, the displacement step of the camera is constant. We think that results should be better if a variable size step is used. To do this, each camera's displacement will be the sum of two vectors:

- A variable size vector $\vec{d}$, starting at the current camera position and whose size and direction are the previous displacement step size and direction.

- A small vector $\vec{\Delta d}$, with $\|\vec{\Delta d}\| \leq \|\vec{d}\|$, whose direction is one of the 8 possible directions (Fig. 4.4), according to the evaluation of the corresponding position using the good view direction criterion.

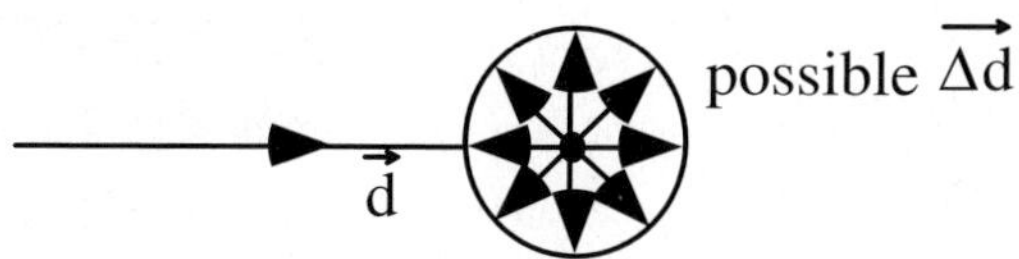

Fig. 4.4: Possible values of $\vec{\Delta d}$

The new position of the camera will be defined by a displacement from its old position according to the vector $\vec{d'} = \vec{d} + \vec{\Delta d}$ (Fig. 4.5).

With this method, the movement of the virtual camera would be slower for regions containing several good view directions, because the displacement step should be smaller.

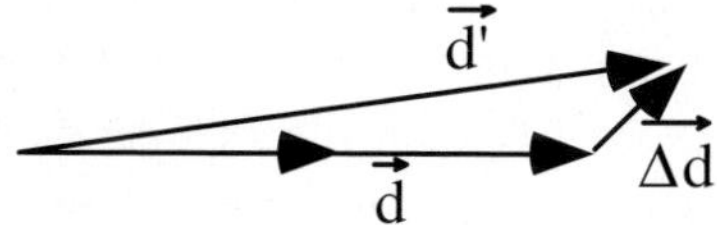

Fig. 4.5: Computing new direction $\vec{d'}$

4.2.2 Fast computation of good view directions

The problem with automatic computation of good view directions is that it is a time consuming process, hardly compatible with a real time smooth movement of a camera. So, in order to reduce the time cost of this task, a computation technique described in chapter 3, using image analysis, is applied.

Based on the use of the OpenGL graphical library and its integrated z-buffer, the technique used is briefly described below.

If a distinct color is given to each surface of the scene, displaying the scene using OpenGL allows to obtain a histogram (Fig. 4.6) which gives information on the number of displayed colors and the ratio of the image space occupied by each color.

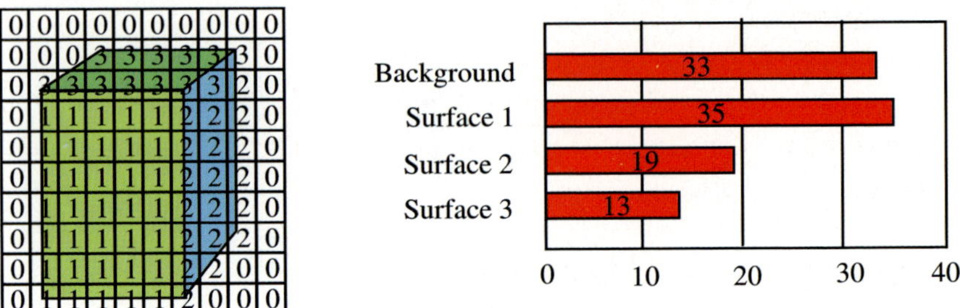

Fig. 4.6: Fast computation of number of visible surfaces and area of projected visual part of the scene by image analysis

As each surface has a distinct color, the number of displayed colors is the number of visible surfaces of the scene from the current position of the camera. The ratio of the image space occupied by a color is the area of the projection of the visual part of the corresponding surface. The sum of these ratios is the projected area of the visible part of the scene. With this technique, the two good view criteria are computed directly by means of an integrated fast display method.

The importance of a viewpoint is now computed by the following formula, first proposed in [13]:

$$I(V) = \frac{\sum_{i=1}^{n} [\frac{P_i(V)}{P_i(V)+1}]}{n} + \frac{\sum_{i=1}^{n} P_i(V)}{r}$$

where: I(V) is the importance of the view point V,

Pi(V) is the number of pixels corresponding to the polygon number i in the image obtained from the view point V,

r is the total number of pixels of the image (resolution of the image),

n is the total number of polygons of the scene.

In this formula, [a] denotes the smallest integer, greater than or equal to a.

The main advantages of this technique are the following:

- Approximated computing of the number of visible surfaces and of the projected area of each visible surface by image analysis is very easy. The total time cost of the technique is $O(d) + O(m+n)$, where $O(d)$ is the image computing cost, m is the number of pixels of the image and n the number of polygons (surfaces) of the scene.

- The display cost with a hardware acceleration based z-buffer is not important and a big number of polygons can be displayed very quickly.

The main drawback of the used acceleration technique is that, with the used OpenGL mode, the maximum number of colors is 4096. Thus a maximum of only 4095 distinct surfaces can be processed. However, sufficiently complex scenes can be constructed with 4095 surfaces. Moreover, with another color encoding model, this limit should be suppressed and the maximum number of colors should be of the order of millions.

4.2.3 Camera movement heuristics

A virtual camera could be defined as a quadruple $(p, \vec{m}, \vec{v}, \vec{b})$ where p is the position of the objective of the camera, $\vec{m}$ the displacement direction, $\vec{v}$ the view direction and $\vec{b}$ the upright direction of a virtual cameraman.

The purpose of the virtual camera movement around the scene is to give the user a good knowledge of the scene properties. To do this, a maximum of interesting regions of the scene must be seen by the camera, with a minimum displacement from the starting point.

In order to avoid a fast return of the camera to the starting point, because of attraction due to the fact that this point probably determines the best view direction, a weight is assigned to each position of the camera. The weight of a position is proportional to the distance of the virtual camera from the starting point. What is the distance of a point on the surface of a sphere from the starting point? We

define this distance as the length of the arc of the circle obtained by the intersection of the sphere by the plane defined by the starting point, the current position of the camera and the center of the scene.

In fact, we must consider two kinds of distance.

- The path traced by the camera's movement (Fig. 4.7a). The length of this path can be computed as the sum of the chords of elementary arcs obtained by decomposition of the path (Fig. 4.7b).

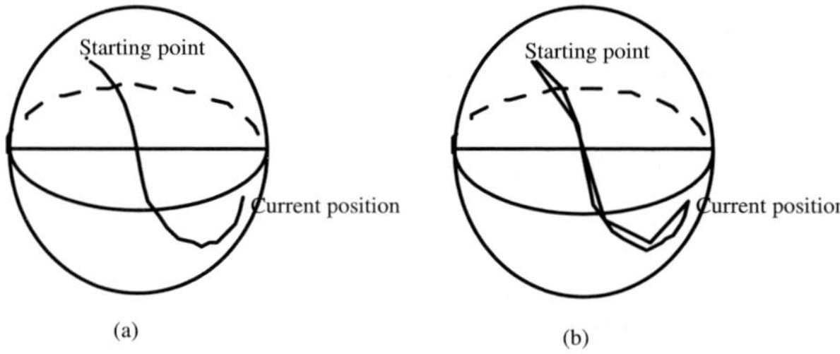

Fig. 4.7: Path traced by the camera and approximated length of the path.

- The minimal length arc between the starting point and the current position of the virtual camera, that is, the distance of the current position from the starting point (Fig. 4.8).

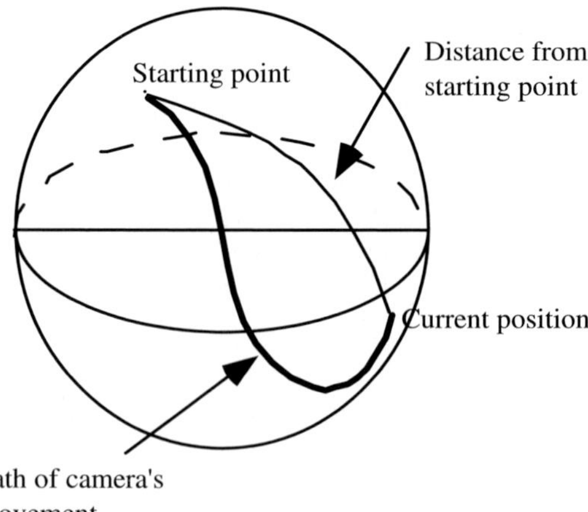

Fig. 4.8: Distance of the current position of the camera from the starting point.

In order to create an heuristic function guiding the movement of the camera, we can observe that the importance of the camera's distance from the starting point is inversely proportional to the length of the path traced by the camera.

Thus, our heuristic function computing the weight of a position for the camera on the surface of the sphere must take into account:

- The global view direction evaluation of the camera's position (n_c).
- The path traced by the camera from the starting point to the current position (p_c).
- The distance of the current position from the starting point (d_c).

Finally, the function we have chosen is the following:

$$w_c = \frac{n_c}{2}\left(1+\frac{d_c}{p_c}\right)$$

where w denotes the weight and c the current position of the camera.

Fig. 4.9 shows incremental global exploration of an office scene, using the method described in this section.

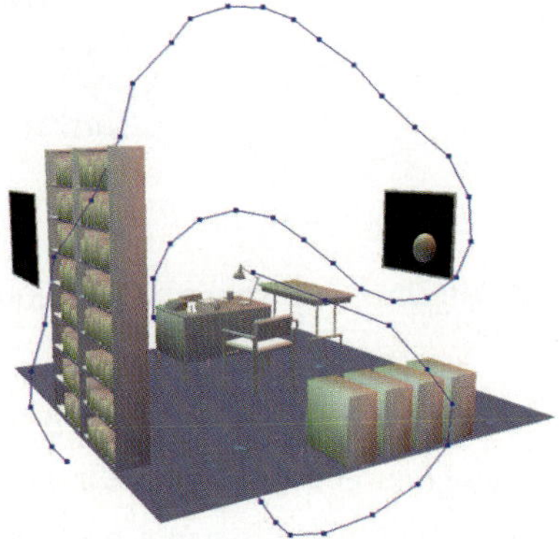

Fig. 4.9: Incremental global exploration of an office (Image G. Dorme)

4.3 Viewpoint entropy-based online exploration

In [5, 6] and [7], the authors propose two methods of virtual world exploration: an outside (global) and an indoor (local) exploration method. The two methods are based on the notion of viewpoint entropy.

The *outside exploration method* is inspired from the incremental outside exploration described in the previous section 4.2. The camera is initially placed at a random position on the surface of a bounding sphere of the world to explore. An initial direction is randomly chosen for the camera. Then, the next position is computed in an incremental manner. Like in the method described in section 4.2, only three possible next position are tested, in order to insure smooth movement of the camera (Fig. 4.10).

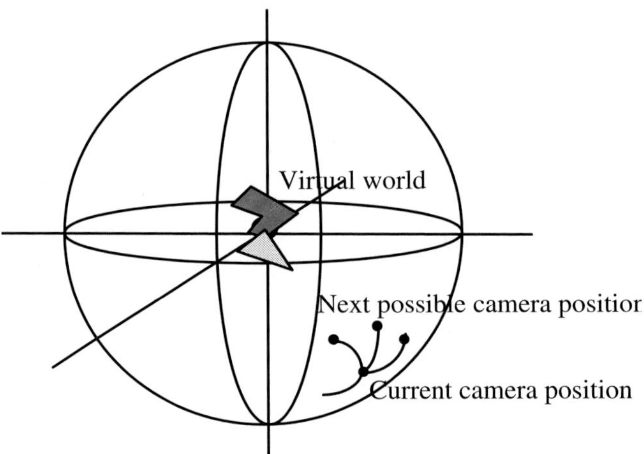

Fig. 4.10: three possible next positions evaluated

An evaluation function computes the best next position of the camera, taking into account only the viewpoint entropy of the *not yet visited polygons* of the virtual world. The exploration is finished when a given ratio of polygons have been visited. The retained criterion of camera position quality seems good enough and the presented results seem interesting. In fact, the used camera position quality criterion is almost the same as in the method of section 4.2, based on the number of visible polygons and on the projected visible area. As only the viewpoint entropy of *not yet visited polygons* of the virtual world is taken into account, the problem of fast return of the camera to a good viewpoint is avoided. On the other hand, exploration is finished when all polygons of the virtual world have been seen by the camera, even if the user would like to continue exploration, in order to well understand the world. Fig. 4.11 shows viewpoint entropy-based exploration of a glass.

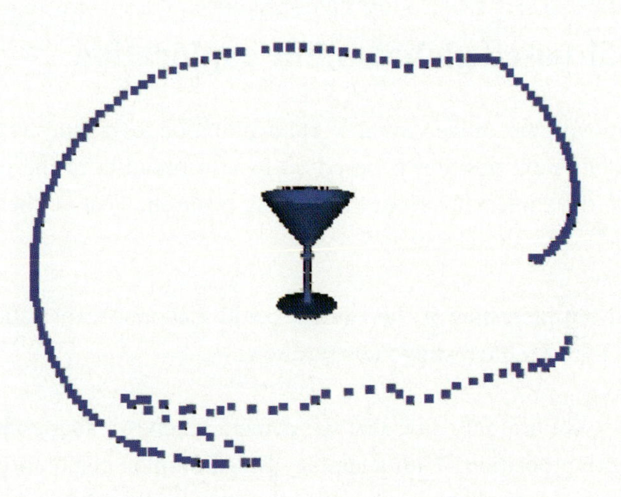

Fig. 4.11: Viewpoint entropy-based exploration (Image P.P. Vazquez)

The *indoor exploration method* is also an incremental exploration technique. The camera is initially placed inside the virtual world to explore and it is supposed to move in this closed environment using each time one of the three following possible movements: *move forward*, *turn left* and *turn right*. Moreover, the camera is restricted to be at a constant height from the floor.

During the virtual world exploration, the path of the camera is guided by the viewpoint entropy of the polygons not yet visited. The next movement of the camera, selected from all the possible movements, is the one with maximum viewpoint entropy coming from the not yet visited polygons. In order to know, at each moment the polygons not yet visited, an array is used to encode the already visited polygons.

The exploring algorithm is stopped when an important part of the polygons of the virtual world has been visited. This threshold value is estimated by the authors to be of about 80% or 90% of the polygons of the explored world.

The main drawback of the method is that collisions are taken into account very roughly by the evaluation function of the camera movement.

4.4   Goal-based global online exploration

The problem with the above virtual world exploration techniques is that the choice of the camera's next position is based on local estimation of the position, trying only to push the camera far from the starting position. That is, there is not a real virtual world exploration strategy.

It would be interesting if the camera could elaborate exploration plans, with the purpose to reach interesting view positions.

Currently, we are only sure that the camera reaches a single interesting position, the starting position. With a strategy based on plan elaboration, reaching interesting positions should be considered as a permanent goal.

4.4.1 Main principles of goal-based exploration

As a partial goal of virtual world exploration could be to reach an interesting position for the camera, it is possible to imagine three different plans [20]:

1. Direct movement of the camera to the next interesting point to reach, taking into account the general purpose rules on the camera's movement.

2. Progressive movement of the camera towards the next interesting point to reach, taking into account the quality of the possible next positions of the camera.

3. Progressive movement of the camera towards the interesting point to reach, taking into account the quality of the possible next positions of the camera and the quality of the possible previous positions before reaching the goal position.

In Fig. 4.12, P1 is the current position of the camera and P2 is an interesting position to reach. The first possibility is to choose the shortest path from P1 to P2, in order to reach P2 as soon as possible. The second possibility tries to reach P2 but the evaluation function determining the next position of the camera takes into account the quality as a point of view of every possible next position. So, the path of the camera is not always the shortest one from P1 to P2. Fig. 4.12 illustrates above all the third possibility: here, the evaluation function is used to determine the next position of the camera, whose current position is P1, and the previous position before the goal position P2, taking into account the quality as a point of view of each candidate position. If the determined next and previous positions are respectively P3 and P4, the same process is applied to determine a path for the camera, where P3 is its current position and P4 the position to reach.

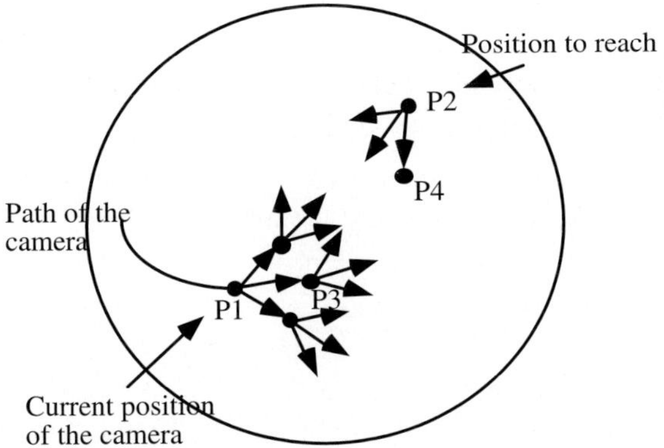

Fig. 4.12: plan elaboration using heuristic search from the starting position to the goal position

This determination of next and previous positions can be made using a heuristic search at one level of depth or at many levels. The evaluation function computing the importance of a position must take into account, at each step, the current distance between the starting position and the position to reach.

So, the proposed goal-based strategy is made of two steps:

- Selection of a small number of interesting position for the camera. This number has to be small in order to allow on-line exploration.

- Virtual world exploration, using one of the three plans proposed above.

During exploration, only not yet seen polygons are taken into account to determine the quality of a camera position.

How to choose a small set of interesting positions to reach on the surface of the bounding sphere? An interesting position is a position being a good point of view. The chosen solution is to divide the bounding sphere in 8 spherical triangles and to compute a good point of view for each spherical triangle, using the technique described in chapter 3.

Now, given the current position of the camera and an interesting position to reach, the following strategy is used when applying the third, most general, exploration plan presented above. If I_p denotes the interest of the position P, $n(P)$ the next position from position P, $p(P)$ the previous position from the position P to reach, pv_p the quality as a point of view of the position P, $d(P1, P2)$ the distance between positions P1 and P2 on the surface of the sphere and Pc and Pr respectively the current position and the position to reach, the next position of the camera can be evaluated using the following formula:

$$I_{n(P_c)} = (pv_{n(P_c)} + pv_{p(P_r)}) \frac{d(P_c, P_r)}{d(n(P_c), p(P_r))}$$

In this formula, the next position of the camera is chosen in order to have the following conditions verified:

- The next camera's position is a good point of view;
- There exists a position, before the position to reach, which is a good point of view;
- The distance between the next position of the camera and the position before the position to reach is minimal.

It is possible to transform the above formula in order to apply the second exploration plan presented at the beginning of this section. Here, only the next position of the camera has to be computed, the position to reach remaining the same (P2). The new formula will be the following:

$$I_{n(P_c)} = (pv_{n(P_c)} + pv_{p(P_2)}) \frac{d(P_c, P_2)}{d(n(P_c), P_2)}$$

The main advantage of the second exploration plan is that exploration is faster. On the other hand, the control of the camera trajectory is less precise than in third exploration plan.

The quality pv_p as a point of view of a given camera position P is computed by using the same evaluation function as in section 4.2.

4.4.2 Implementation of exploration techniques and results

The first exploration plan described in subsection 4.4.1 was implemented but it is not very interesting, as smooth movement of the camera is not guarantied. The second exploration plan produces interesting results, which will be commented in this section. The third exploration strategy is not yet implemented.

We have chosen to comment results of the proposed plan-based method, obtained with two scenes: The office scene, already seen in section 4.2 (Fig.4.9), and a sphere with 6 holes, containing various objects inside. This second test scene is very interesting because most of the scene details are visible only through the holes.

Four cases were studied, corresponding to variants of the second formula of subsection 4.4.1:

1. Only the distance between the current position of the camera and the position to reach is considered. The viewpoint qualities of positions to reach are not taken into account (Fig. 4.13).

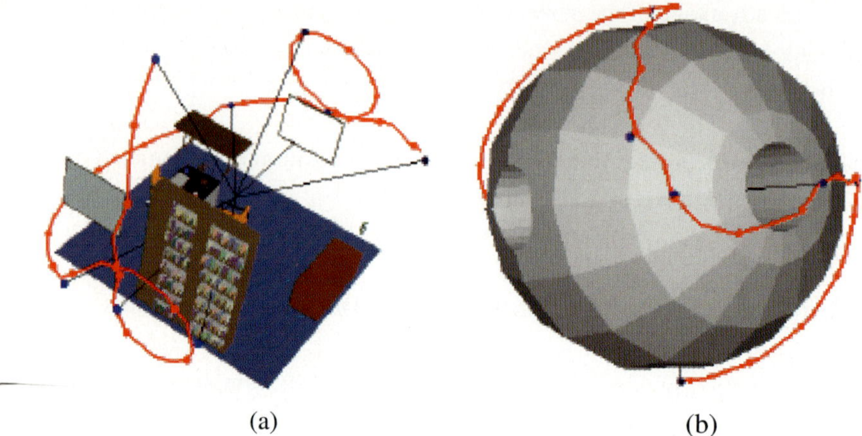

(a)   (b)

Fig. 4.13: Only distance to the next position to reach is taken into account (Images S. Desroche)

2. Only the viewpoint qualities of positions to read are considered, that is, the next point to reach is the closest to the current camera position (Fig. 4.14).

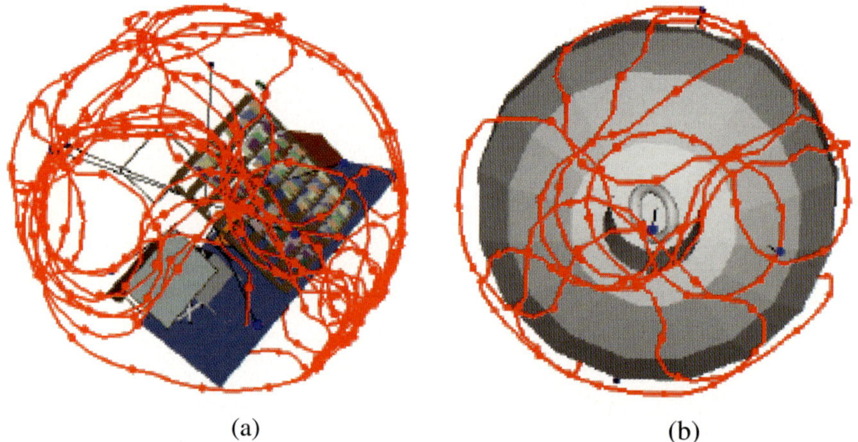

(a)   (b)

Fig. 4.14: Only viewpoint qualities of the positions to reach is taken into account (Images S. Desroche)

3. Both distance and viewpoint quality of the position to reach are considered, with equal weights (ratio viewpoint quality/distance = 1). See Fig. 4.15.

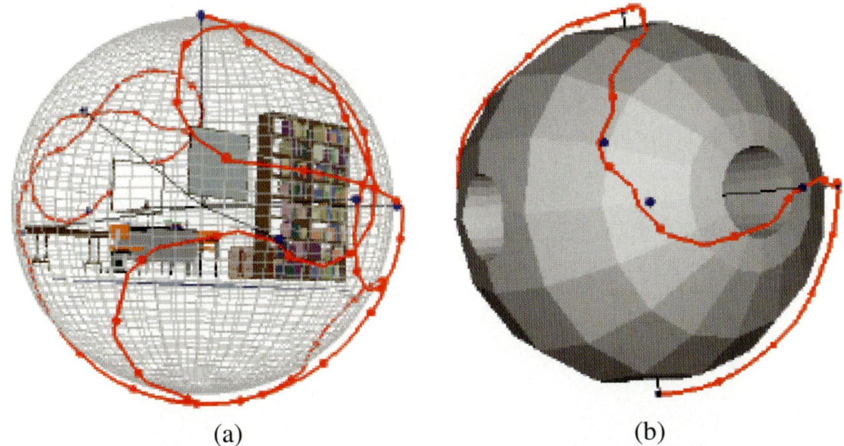

(a)   (b)

Fig. 4.15: Both distance and viewpoint quality are equally considered (Images S. Desroche)

4. Both distance and viewpoint quality of the position to reach are considered, with the distance weight equal to twice the viewpoint quality weight (ratio viewpoint quality/distance = 1/2). See Fig. 4.16.

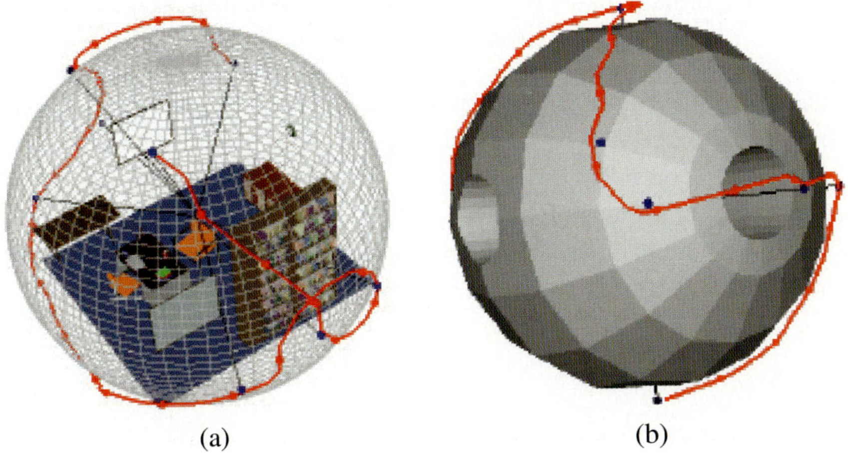

(a)   (b)

Fig. 4.16: The distance is twice more important than viewpoint quality (Images S. Desroche)

One can see that the fourth case gives the best results with both test scenes. With the sphere test scene, the fourth case gives results very close to those of the first case (distance only), as it can be seen in Fig. 4.17.

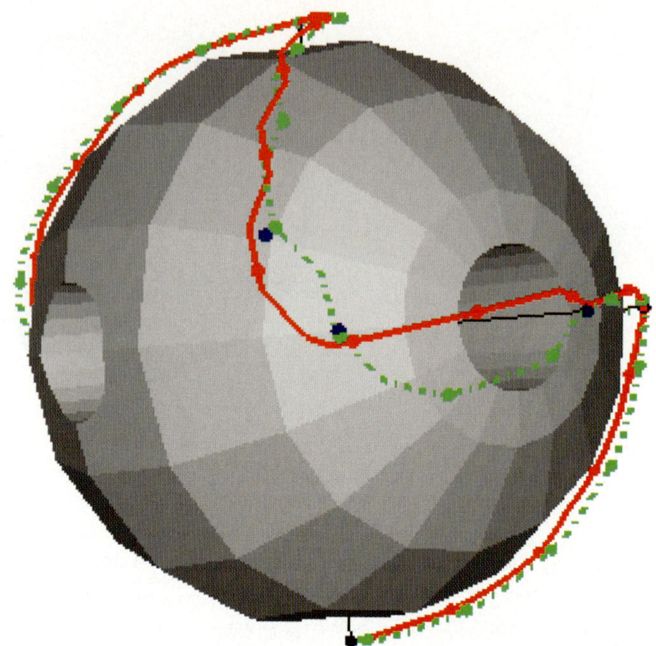

Fig. 4.17: The distance-based path (in green) is very close to the best path (in red)
(Image S. Desroche).

Graphic representations of results (Fig. 4.18 and Fig. 4.19) show that, in all cases, more than 80% of the scene is seen when the last of the initially defined positions is reached by the camera. This proves that the chosen method for creating the small initial set of positions to reach is well adapted to the purpose of the plan-based automatic exploration. The number of positions to reach of the initial set is generally small (usually 8), so that the added extra time cost is negligible.

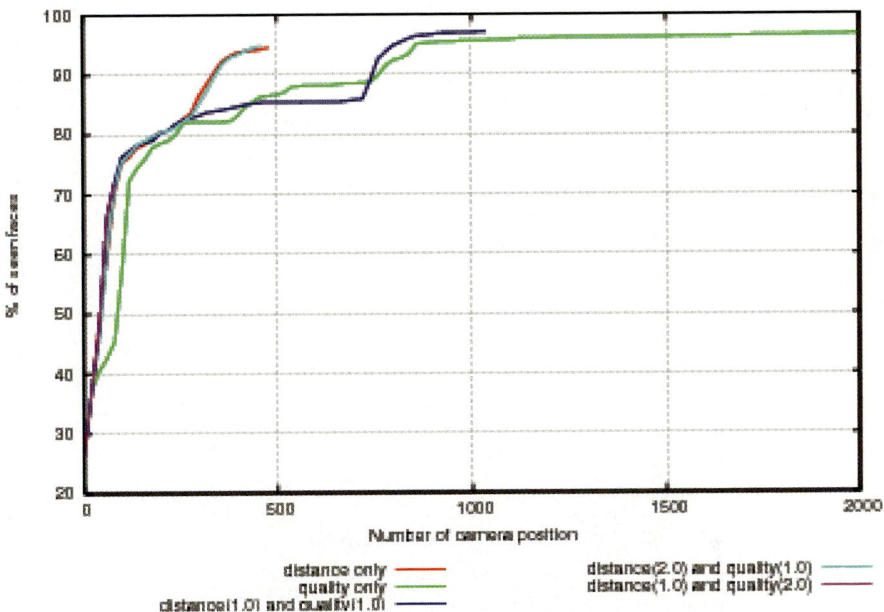

Fig. 4.18: Graphic representation of results with the Office test virtual world

4.4.3 Discussion

The technique presented in this section performs incremental determination of the next position of the camera. It takes into account both viewpoint quality and positions that should be reached by the camera and tries to elaborate exploration plans in order to reach these positions when it tries to determine the next camera position.

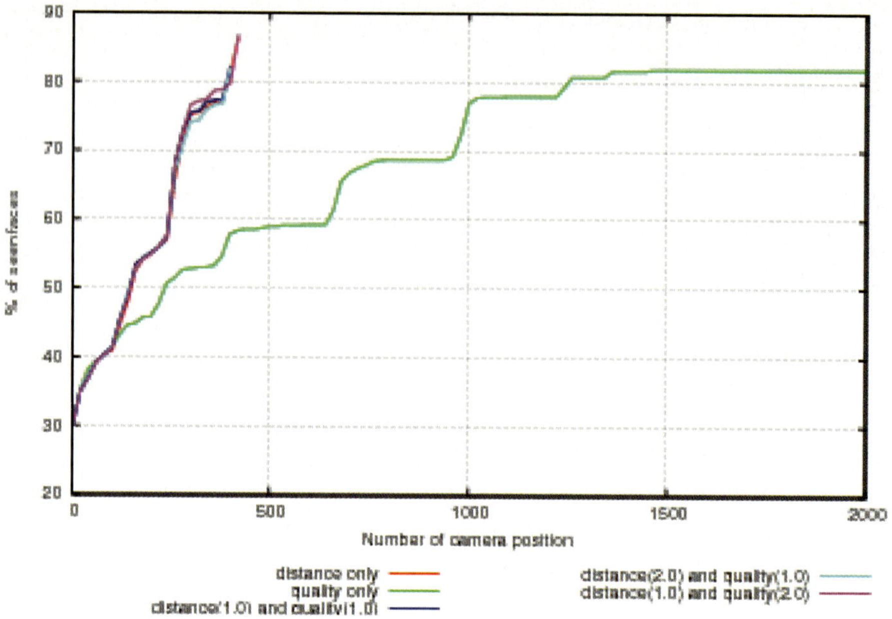

Fig. 4.19: Graphic representation of results with the Sphere test virtual world

The study of obtained results shows that the distance of the current camera position from the position to reach is often more important than the viewpoint quality of the current position. In our opinion this fact can be explained by the mode of choice of the initial set of positions to reach. As they are chosen to be good viewpoints and well distributed on the surface of the surrounding sphere, it is natural to discover a big part of the virtual world by visiting the positions to reach only according to their distance from the current position. The optimal ratio viewpoint quality/distance is 1/2. Obtained trajectories with this ratio are always interesting and the used methods are not time consuming. In implemented techniques, the movement of the camera is smooth and the chosen camera paths interesting.

Taking into account that the implemented goal-based technique uses the second of the three possible camera's movement plans, one can hope that implementation of the more accurate third plan will give much more interesting results. It would be interesting to find an objective measure in order to evaluate the different scene exploration techniques.

4.5 Local incremental online exploration

Unfortunately, even if global virtual world exploration gives interesting results, it is difficult to apply to some scenes comporting parts difficult or impossible to see by a camera moving outside the scene (Fig. 4.20). This is the reason for which another family of methods is proposed, so-called *local virtual world exploration methods*, where the camera moves inside the virtual world to explore, in order to have a closer view of some elements composing the scene.

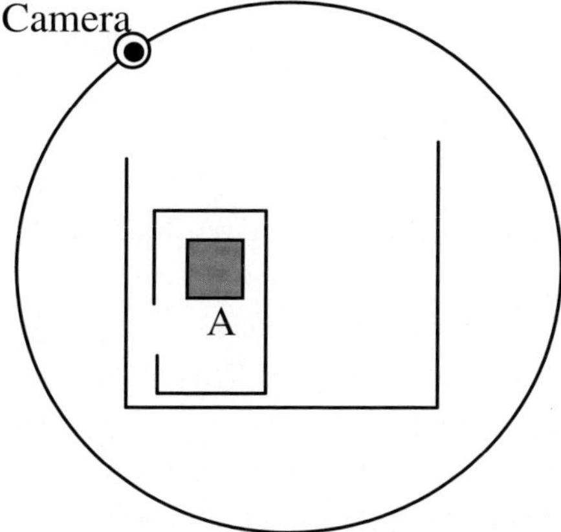

Fig. 4.20: The object A is not visible by a camera moving outside the scene

The space of movement of the virtual camera is defined by a bounding box of the scene to explore. In order to have a finite number of possibilities, this search space is sampled and the problem to resolve is the passage from the current discrete point of the search space to the next one.

The goal of the method is to explore the scene as well as possible with a minimum movement of the virtual camera [3, 4]. This movement must take into account:

- **Possible obstacles**, The movement of the virtual camera must avoid the neighboring of obstacles.

- **Points of the scene which must be visited**. If the user desires that the camera visits some positions in the scene or if we are able to determine automatically the positions in the scene that seem interesting to visit, these positions must be reached by the camera as soon as possible.

- **Quality of view from the current position of the camera**. The quality of view from the current position is a very important criterion because the aim of the camera's movement is to understand the scene as well as possible.

The method presented here and its current implementation does not use the last of the above criteria. It will be taken into account in subsequent work and publications.

Before explaining the technique used to guide the movement of the camera inside the scene, we must fix some limits to the manner the virtual camera is handled.

Usually, a camera has 6 degrees of liberty: its position (3 degrees), its direction of view (2 degrees) and its angle of rotation from the view direction axis (1 degree).

We have decided to simplify the problem by supposing that the camera is always oriented towards its next position. This restriction avoids movements like travellings but this is not the object of this work. The hypothesis that the camera is oriented towards its next position requires to always compute not only the next position of the camera but also the position that will follow the next position.

The last degree of liberty, the angle of rotation from the axis of the direction of view, will get an arbitrary value in order, for the user, to have a fixed notion of "up" and "down".

4.5.1 Preprocessing: analysis of the scene

Even if this local exploration method is presented as an incremental online method, before the beginning of the camera movement inside the scene, a preprocessing phase is required, based on a semantic analysis of the scene.

The preprocessing phase works as following:

- The workspace is sampled and represented by a matrix of voxels.
- A weight is assigned to each voxel (Fig. 4.21) according to:
    - The nature of the voxel (occupied or not by an element of the virtual world).
  - The distance (length of the shortest path) to a point to be visited.
  - The distance from an obstacle.

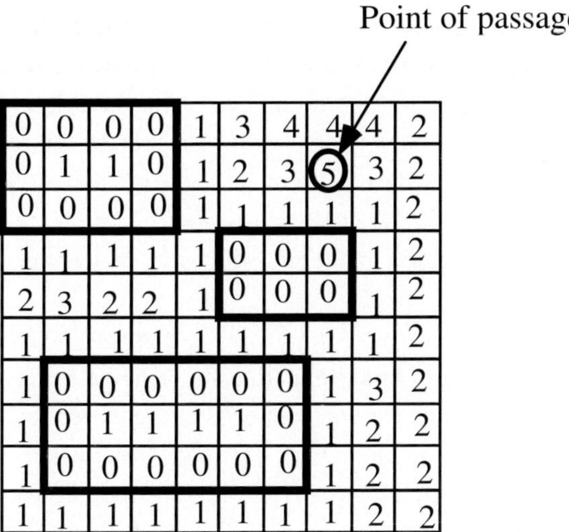

Fig. 4.21: Sampled workspace and weights assigned to each point

In this implementation, the weight assigned to a voxel is a triple corresponding to the above criteria.

In a first implementation of the method, distances, used in the second and third criteria, were Manhattan distances. So, for two given voxels of coordinates (x1, y1, z1) and (x2, y2, z2) respectively, their Manhattan distance is equal to:

$$d = |x_2-x_1|+|y_2-y_1|+|z_2-z_1|$$

In the current version of the program, the used distances are Borgefors distances [ ].

The analysis phase, which assigns a weight to each voxel of the sampled virtual world, works according to the following algorithm:

1. Fill the forbidden cells of the matrix (cells occupied by objects of the scene) with the worst (minimum) value (for example 0).

b. Fill the other cells of the matrix, starting from the obstacles (objects of the scene) and filling each time the adjacent cells of the current cell with the value v+d, were v is the value of the current cell and d the distance from the current cell. In case of conflict between the value assigned to the cell and its previous value, the lower value is chosen.

c. Fill the cells of the matrix belonging to points of passage with a maximum value M.

d. Fill the cells of the matrix, starting from the cells belonging to points of passage and filling each time the adjacent cells of the current cell with the value v-d, were v is the value of the current cell and d the distance from the current cell. The filling is continued until expansion has reached the limit of the influence field of points of passage. In case of conflict between the value assigned to the cell and its previous value, the greater value is chosen.

The cells of the matrix can be filled according to the 6 main directions. Other filling methods, based on the work of Rosenfeld-Pfalz or on the work of Borgefors [Bor 86] can also be used.

At the end of the filling phase, each cell of the matrix contains at least three kinds of information: the state of the cell (occupied or not), its value due to the filling started from obstacles and its value due to the filling started from points of passage.

In the case where no points of passage are given, it is possible to automatically compute some interesting positions by choosing an arbitrary number of cells corresponding to local maximums according to their distance from obstacles. This choice insures a wider view of the scene from the selected cells.

4.5.2 Determining the trajectory of the camera

In order to explore a scene by a virtual camera which moves inside it, two kinds of information are needed: The position of the starting point of the camera and the next position of the camera, computed from its current position and its current direction of movement. This information is given by an evaluation function.

The evaluation function, which computes the next position of the camera, takes into account the weight of a voxel and other heuristics like the path traced by the camera compared to the distance of the current position from the starting point or the direction of the camera movement.

The quality of view from a position of the camera is not taken into account in the current implementation. It can be computed for some voxels of the workspace and estimated for all the others.

The weight $w(V)$ of a voxel V is computed by taking into account the following considerations: we think that the information concerning the attraction force of a point of passage is more important than the distance from obstacles. So, a voxel belonging to the influence field of a point of passage has a priority greater than another not submitted to the influence of any point of passage. This is done in the following manner for a voxel V represented by the triple (s, v_o, v_p), where s is the state of the voxel (occupied or not), v_o its value due to the filling started from obstacles and v_p its value due to the filling started from points of passage:

If $v_p > 0$, that is if the voxel V is in the influence field of a point of passage, $w(V)=D+v_p$, where D is the maximum distance in the matrix. Otherwise, $w(V)=v_o$.

The starting point of the camera's movement is one of the border voxels of the matrix, chosen according to its weight. The starting position of the camera being determined, its next position is computed by choosing the neighboring voxel with the best (greatest) weight. The knowledge of these two positions, initial and next, determines a direction, the current movement direction of the camera. When the current position of the camera is the starting point, its next position is chosen among all the neighboring voxels (17 possibilities) whereas in all other cases the next camera's position is chosen by taking into account the direction of the camera movement: in order to insure smooth camera's movement, only the neighboring positions corresponding to a variation of direction lower than or equal to 45 degrees (9 possibilities) are considered for computing the next position of the camera. Evaluation of a candidate next position can be made at any level of depth, depending on the time available.

In case of conflict (two or more directions of the camera have the same value), the next position of the camera is chosen randomly. The main advantage of this solution is that the scene exploring process is a non deterministic one and avoids systematic errors. Another possibility could be to do a new evaluation of the conflicting positions with one more level of depth.

4.5.3 Processing of a point of passage

The local virtual world exploration method presented in this section, insures that, if the current position of the camera is under the influence of a point of passage, this point will be visited by the shortest possible path, taking into account obstacles that possibly separate the point of passage from the current camera's position.

Whenever a point of passage has been visited by the camera, this point is no more considered as a point of passage, in order to avoid the camera to remain in the neighboring of this point and to permit it to reach another point of passage, the closest to its current position. So, the values due to the filling started from points of passage of all the cells being in the influence field of the current voxel have to be updated.

4.5.4 Discussion

The method of local incremental online virtual world exploration presented in this section, based in the camera's movement inside the scene, has been implemented but results are not entirely satisfactory.

In Fig. 4.22, one can see the path of a camera exploring the inside of a seashell. One single point of passage was manually defined near the end of the spiral. The camera goes to this point as straight as possible.

In Fig. 4.23, the office scene is locally explored by a virtual camera. 20 points of passage were randomly chosen. Exploration cannot reach all these points of passage.

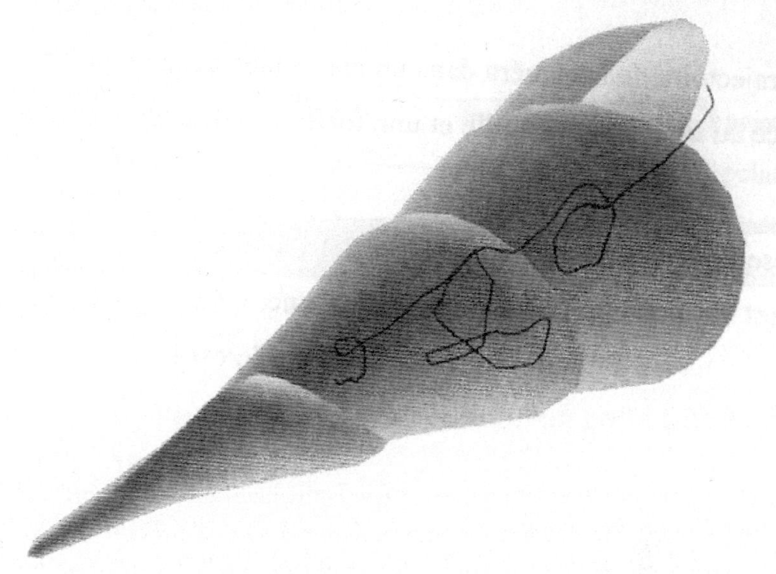

Fig. 4.22: Local exploration of a seashell (Image G. Dorme)

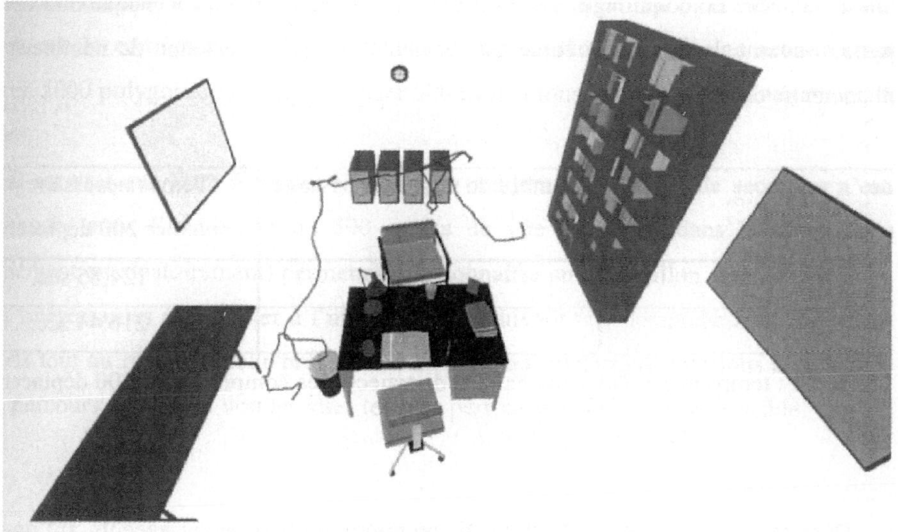

Fig. 4.23: Local exploration of an office (Image G. Dorme)

4.6   Minimal viewpoint set-based global offline exploration

This section presents new techniques and methods proposed for automatic off-line exploration of virtual worlds. The exploration process is generally performed in two steps [11, 12]. The goal of the first step is to find a set of points of view allowing to see all the details of the scene. In the second step, the selected points are used in order to compute an optimal trajectory. As it has been explained above, exploration is a global one and the selected points of view are computed on the surface of a sphere surrounding the virtual world.

4.6.1 Computing an initial set of points of view

Computing an initial set of points of view is the main part of our off-line exploration techniques. The final set should be minimal and should present as many information of the scene as possible. As an object may be seen from one or more viewpoints, we will have to avoid redundancies in order to keep only useful viewpoints.

The first thing to do is to build a beginning balanced set of viewpoints placed on a sphere which surrounds the scene.

The center of the surrounding sphere is the center of the scene and its ray is calculated with the following empiric formula:

$$\rho = 1.5 * \sqrt{\frac{(Xmax - Xmin)^2 + (Ymax - Ymin)^2 + (Zmax - Zmin)^2}{3}}$$

Viewpoints distribution depends on the increment of two loops:

> **For** φ *from 0 to* 2π *do*
>   **For** θ *from 0 to* π *do*
>   *Point* (ρ, φ, θ) *is an initial viewpoint*
>   **End**
> **End**

Smaller increments for φ and θ give a better quality of the beginning set. In Fig. 4.24, one can see some examples with different increments.

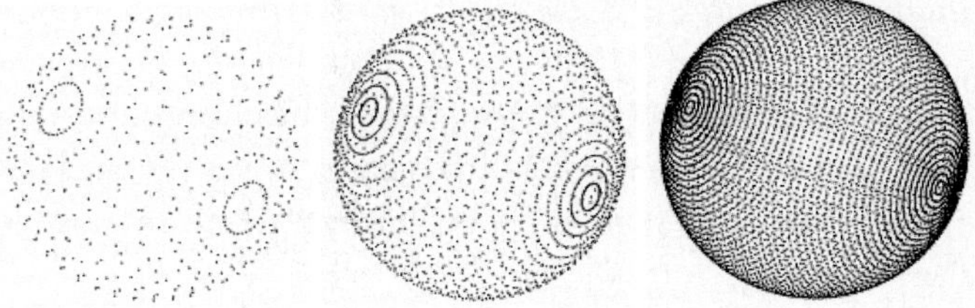

Fig. 4.24: Starting set of viewpoints (Image B. Jaubert)

Three criteria have been defined in order to select the most interesting points of view. The first one, as commonly used in exploration methods, is related to the number of visible polygons. The second criterion depends on the number of visible *objects*. The third criterion depends on the above ones and on the angle of each face with the view direction.

Visible polygons

This kind of evaluation considers the number of visible polygons of each point of view. Intuitively, we assume that a point of view is more interesting if it allows to see more polygons. Unfortunately, if this is true in many cases, for some scenes this criterion is not sufficient because the number of independent parts in the scene is at least as important as the number of polygons. However, the visible polygons criterion will be used, combined with other criteria.

Visible objects

In order to resolve the problem due to some limits of a face-based exploration the concept of object is considered with no consideration of the number of its faces. Indeed, the most interesting point of view allows perceiving the maximum of objects of the scene. This is the natural human perception: in order to describe a room, we first describe the objects it contains. The perception of these objects needs only a partial vision of them. Therefore, it is obvious to consider this criterion as the main one.

Combining evaluation criteria

A third criterion is added to previous criteria, which takes into account not only the number of visible polygons but also the quality of view (that is the angle of vision) of a polygon from the point of view. This criterion is resumed in the following formula.

$$mark = (p \times o) + \sum_{polygons} (p_v \times \cos(\alpha))$$

In this formula, p is a real number and represents the number of polygons in the scene divided by the number of objects, o is the number of objects in the scene, p_v is a value between 0 and 1 and shows the visible part of the polygon (0.5 means that the face is 50% visible) and α is the angle between the polygon orientation and the vision direction (Fig. 4.25).

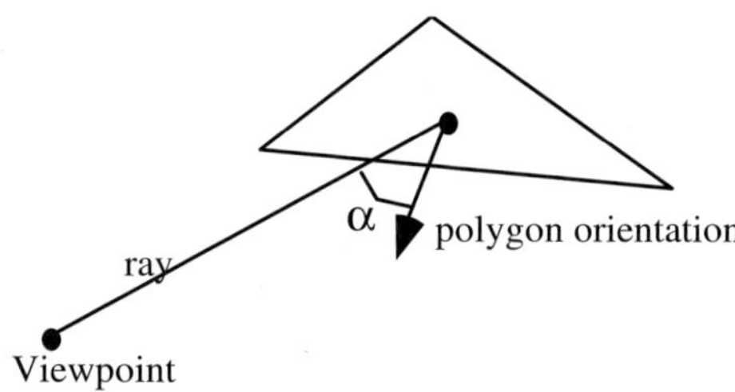

Fig. 4.25: Angle of vision

This criterion offers a additional evaluation mode for a viewpoint using all previous criteria together with quality of view.

1. Point of view evaluation

The point of view evaluation depends on the values returned by the three criteria which are sorted in the following order, from the more important to the less important: *number of visible objects, number of visible faces* and *combination of the above criteria*. For example, let's consider these three criteria. If two different points of view perceive eight objects of the scene, the best one would be the point which sees the greatest number of faces. If the number of visible faces for the different objects is the same, the best point of view will be determined by the third criterion. If the values of the three criteria are equal for the two points, one of them will be randomly chosen as the best viewpoint.

2. Point of view selection

Evaluation function

The goal of an evaluation function is to estimate the view quality from a point, in order to choose the most interesting viewpoints. This is very important during the exploration procedure. This function returns a point *value* using various criteria. Viewpoint evaluation uses two main tools.

The first tool uses an intersection determining function. This tool says whether the intersection of the polygon with a ray defined by the point of view and a point on a polygon is visible (Fig. 4.26).

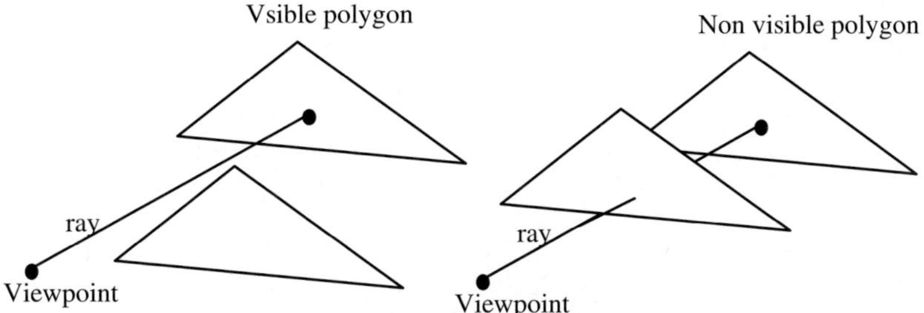

Fig. 4.26: Intersection function

The intersection algorithm is the following:

```
Function Intersection (viewpoint, polygon) : boolean
        result ← true
        ray ← make ray (view point, random polygon point)
        for each other polygon do
            if ray is stopped before the point then
                        result ← false
            end
        end
        return result
```

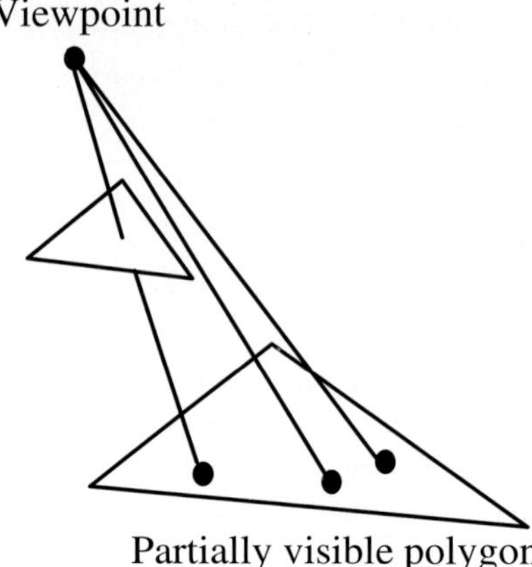

Fig. 4.27: partially visible polygon

The second tool determines the visible part of a polygon from a point of view. To do this, a more or less important number of rays is sent from the point of view to randomly generated points of the polygon (Fig. 4.27). This visible part is given by the following formula.

$$visibility = \frac{number\ of\ non\ stopped\ rays}{number\ of\ rays}$$

The corresponding algorithm is the following:

```
Function visibility (view point, polygon) : real (between 0 and 1)
    Visibility ← 0
    for ray from 1 to Number_of_Rays
        visible ← Intersection (view point, polygon)
        If visible is true then
            visibility ← visibility + (1/ Number_of_Rays)
        end
    end
Return visibility
```

The above tools are used by the global viewpoint evaluation function. This function works in the following manner:

- If a polygon of an object is visible, the object is considered visible.
- At the end of the evaluation process, the function returns three values (seen objects, seen polygons and an evaluation mark of the viewpoint.

The evaluation function algorithm may be schematically presented in the following algorithm, which defines the function *evaluate*.

```
function evaluate (scene, viewpoint) : three real
    for each object in the scene do
        for each polygon of the object do
            visible ← visibility (viewpoint, polygon)
            if visible > 0 then
              If the object is not globally visible
                 object ← locally visible
              end
              If the polygon is not globally visible
                 polygon ← locally visible
                 Add 1 to Seen polygons number
                 mark ← mark + cos (ray, polygon)
            end
          end
        end
        if the object is locally visible
            Add 1 to Seen objects number
        end
    end
    return (Seen objects number, Seen polygons number, mark)
```

In this function, a globally visible polygon or object is permanently marked to indicate that it is visible from another viewpoint.

3. How to mark a polygon or an object as globally visible?

We need a function for marking a polygon or an object permanently to remember which one has already been seen from a selected viewpoint. After the evaluation function finds the best viewpoint, we need to save the set of polygons and objects seen from this viewpoint in order to take them into account. The next iteration of the evaluation function will know which part of the scene was already seen and will choose another viewpoint which fill up the set.

This function works like the evaluation function and marks the globally visible objects and polygons instead of evaluating the viewpoint. The number of rays is much more important than in the evaluation function to assure better results.

```
procedure marking (scene, viewpoint)
    for each object in the scene do
        for each polygon of this object do
            for num from 1 to (number of rays* precision)
                visible ← Intersection (viewpoint, polygon)
                if visible is true then
                    object ← globally visible
                    polygon ← globally visible
                end
            end
        end
    end
```

4. First viewpoint selection method

This method uses the initial set of viewpoints and applies the evaluation function on each of them.

The best viewpoint (according to chosen criteria) of the set is selected and added to the final set of viewpoints.

The mark function marks the objects and polygons seen from selected viewpoints.

The evaluation function is applied on each point of the initial set (except the viewpoints already selected in the final set) and the best viewpoint is selected and added to the final set. The viewpoint evaluation function does not take into account already seen objects and polygons.

The whole process is repeated until the terminal condition is satisfied.

The following algorithmic scheme describes this viewpoint selection method.

```
function first method (scene, initial set of viewpoints): final set of viewpoints
do
        Chosen point ← none
        best note ← ( -∞, -∞, -∞)
        for each viewpoint in the initial set do
                If viewpoint is not in the final set
                        note ← evaluate (scene, viewpoint)
                        if note > best note then
                                Chosen point ← viewpoint
                                best note ← note
                        end
                end
        end
        If Chosen point ≠ none
                mark(scene, chose point)
                add(final set, Chosen point)
        end
until terminal condition or chosen point = none
return (final set)
```

It is easy to see that in this method:

- All the objects and polygons which can be seen are detected and at least one point of view in the final set can see them.
- The final set of viewpoints is minimal.
- The time cost to compute the final set of viewpoints is maximal. The initial set of viewpoints has to be processed n+1 times if n is the number of viewpoints of the final set.

5. Second viewpoint selection method

In this method all the viewpoints of the initial viewpoint set are evaluated and sorted, according to their values, in decreasing order. The viewpoints of the obtained classified viewpoint set may now be processed by order of importance. The best viewpoints will be processed before the others.

```
function Second Method (scene, initial set of viewpoints): final set of viewpoints
sorted array ← empty
Number of points of view ← 0
For each point in the initial set
        evaluation ←  evaluate(scene, viewpoint)
        add in sorted array (viewpoint, evaluation)
end
marking (scene, sorted array[0]) (the best viewpoint)
Select useful viewpoints (scene, sorted array)
return (sorted array)
```

The *Select useful viewpoints* procedure tries to suppress redundant viewpoints.

- The evaluation function is used on the current point (except the first one which is already marked).
- If this point allows to see new polygons or objects, we mark them and go to the next point.
- Otherwise the viewpoint is erased from the sorted array.

```
procedure Select useful viewpoints (scene, array)
    for sign from 1 to number of viewpoint-1 do
        evaluation ← evaluate (scene, array[sign])
        if evaluation ≠ blank evaluation then
            mark (sorted array[sign])
    else
            erase from sorted array (sign)
            Retract 1 from number of viewpoints
        end
    end
return (array)
```

This method has some advantages and some drawbacks, compared to the first method.

- All the objects and polygons which can be seen are detected and at least one point of view in the final set can see them.
- The final set is not minimal and the number of viewpoints in the final set can be much higher than with the first method.
- The method is less time consuming than the first one. The initial set of viewpoints is processed two times.

6. Improvements

It is possible to improve the proposed methods of computing reduced viewpoint sets, in order to reduce computation time or to improve the quality of selected viewpoints.

Bounding boxes of objects may be used to accelerate the whole process. Thus, if the bounding box of an object is not intersected by the current ray, it is not necessary to test intersection of the ray with the corresponding object or with polygons of this object.

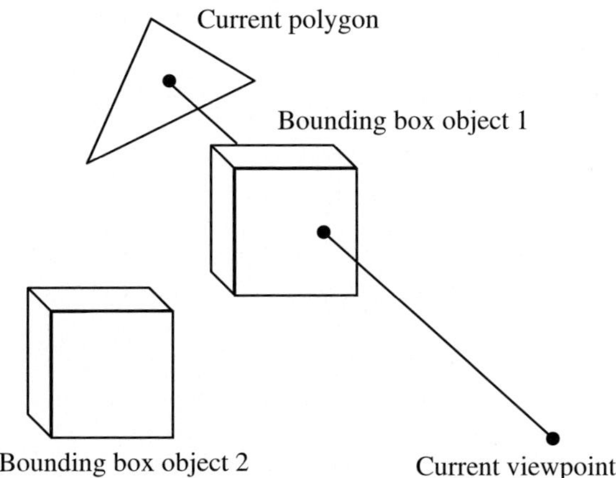

Fig. 4.28: Using bounding boxes to reduce computation cost

In Fig. 4.28 the polygons of object 2 cannot hide the current polygon because the current ray does not intersect its bounding box.

Bounding boxes may also be used instead of objects, in order to accelerate the computation in a first step of the processing where only visible objects will be processed. A second step is then necessary for the non visible objects, in order to verify more precisely some cases difficult to resolve. The use of bounding boxes allows dividing the computation time by a quantity equal to $\frac{average \quad of \quad faces \quad by \quad object}{6}$ but the main drawback of this improvement is that embedded objects will never be marked as visible during the first step of the process.

The quality of selected points of view may be improved in the following manner: For each point of view of the final set, some points in its neighborhood are tested and, if they allow to see new objects or polygons, they are included in the set. This improvement allows to find new potentially interesting viewpoints but the computation time and the number of selected viewpoints increases.

7. Computing an optimized trajectory

After the reduced set of viewpoints is computed, we have to compute an optimized camera trajectory using the points of this set. The proposed method works in two steps. The first step performs an ordering of the viewpoints of the set while the second step computes the final trajectory.

8. *Ordering of viewpoints*

As the camera will have to visit all the viewpoints of the reduced set of viewpoints, an order must be defined for this visit. This order may depend on various criteria.

Proximity-based ordering

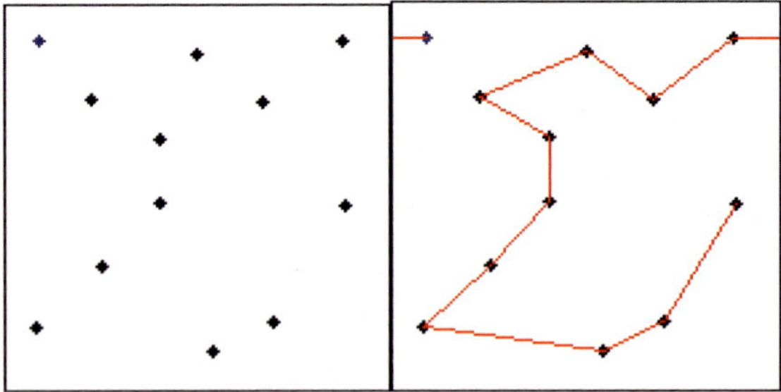

Fig. 4.29: Proximity-based ordering

In this method (Fig. 4.29), proximity of viewpoints is considered as the first ordering criterion:

- The best viewpoint of the reduced viewpoint set is chosen as starting point for the camera trajectory.
- The next trajectory point is the closest to the current point viewpoint. This point becomes the current trajectory point.
- The previous step is repeated for each remaining viewpoint of the reduced viewpoint set.

This ordering is not always satisfactory, especially if the camera uses a straight line trajectory to go from a position to the next one and several trajectory points are close to each other.

Minimal path-based ordering

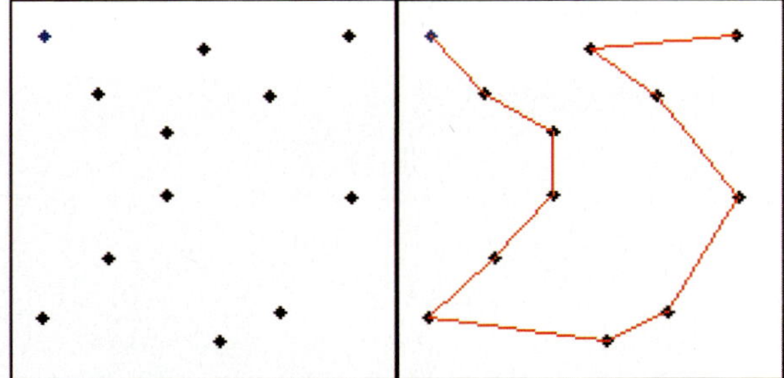

Fig. 4.30: Minimal path-based ordering

In this method (Fig. 4.30), the main ordering criterion is the total length of the whole trajectory, which has to be minimal. This ordering criterion gives better results than the proximity based one but the ordering process is more time consuming. Ordering is achieved by testing all the solutions and retaining the best one. Optimization is not necessary for the moment because the number of points of view is small and exploration is made off-line.

9. Computing the camera trajectory

At the end of the ordering process, applied to the reduced set of viewpoint, it is possible to compute a camera trajectory.

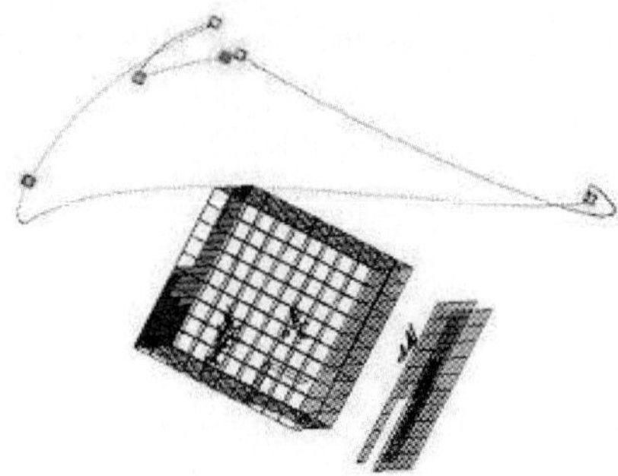

Fig 4.31: Smooth trajectory for scene exploration (Image B. Jaubert)

The method has to avoid abrupt changes of direction in the camera movement. To do this, the following technique is used.

Let us suppose (Fig. 4.33) that the current camera position on the surface of the sphere is A and the viewpoint to reach is B.

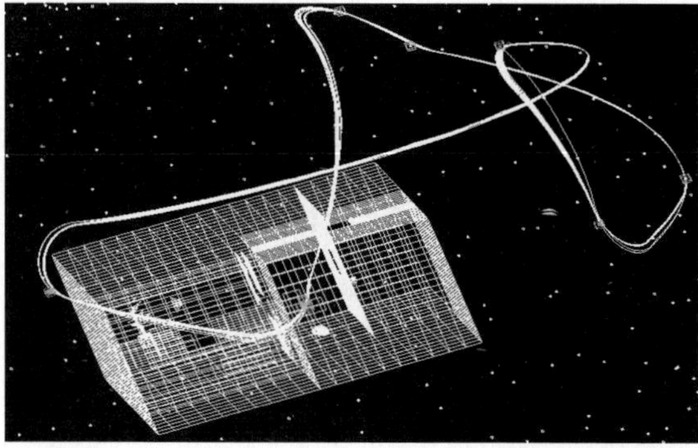

Fig. 4.32: Another example of smooth scene exploration (Image B. Jaubert)

If the current direction of the camera movement at point A is d, a new direction AA1, between d and AB is chosen and the next position A1 of the camera is defined on the vector AA1. The same process is repeated from each new step. In Fig. 4.33, A, A1, A2, … are successive positions of the camera.

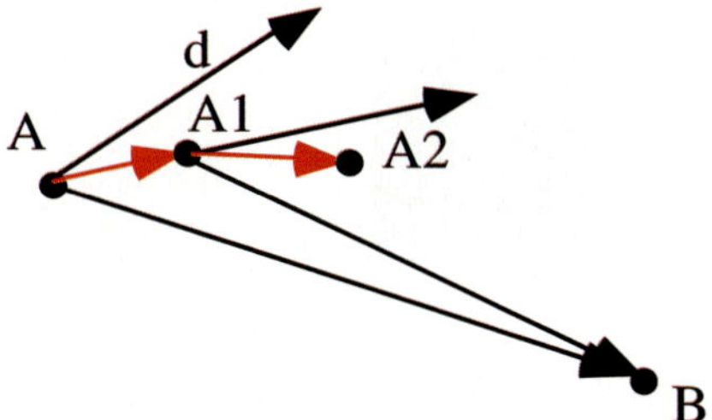

Fig. 4.33: Computing a smooth camera trajectory

In Fig. 4.31 and Fig. 4.32 one can see illustration of smooth camera trajectories obtained with the above technique.

10. Discussion

Performances of the two viewpoint selection methods, presented in subsection 4.6.8, are compared here for the same set of 4 scenes.

Comparison of the proposed methods is made with an initial set of 17 viewpoints. The purpose of these methods is to compute a reduced set of viewpoints. Of course, the minimal set of viewpoints needed to understand a scene is strongly dependent on the nature of the scene. For the scene of Fig. 4.34 a single point of view allows to well understand it whereas for the scene of figure 4.35 at least 4 points of view are needed.

Fig. 4.34: Test scene for a single point of view (Image B. Jaubert)

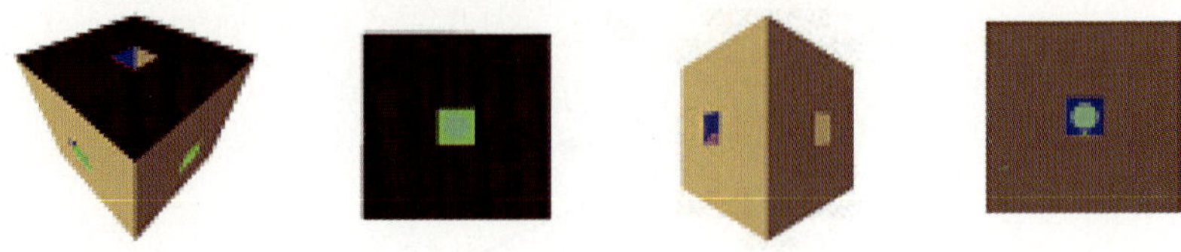

Fig. 4.35: Scene with multiple points of view needed (Images B. Jaubert)

Four different scenes, A, B, C and D, are used for comparisons and two criteria were chosen to evaluate the two methods: number of points of view of the reduced set of viewpoints computed by each method and time cost of each method.

Results of evaluation using the number of points of view criterion are shown in Table 4.1 where one can see that the two methods give very often the same results but the first method is generally better than the second one because the computed set of viewpoints is really minimal.

Table 4.2 shows results of evaluation of the two methods using the time cost criterion.

	Sc A	Sc B	Sc C	Sc D
Method I	4	3	4	3
Method II	4	3	4	6

Table 4.1: Number of computed points of view per method

	Sc A	Sc B	Sc C	Sc D
Method I	1m57s	3m49s	4m18s	5m49s
Method II	1m23s	3m07s	3m19s	3m30s
Gain	29%	18%	22%	39%

Table 4.2: Computation time per method

The second method is always faster than the first one. The obtained gain grows with the complexity of the scene, even if the notion of scene complexity is not formally defined. This complexity is approximated by the number of needed viewpoints for understanding the scene.

Both, proximity-based and minimal path-based viewpoint ordering, methods give satisfactory results.

The techniques presented in this section, especially the ones used to compute a minimum set of points of view, may also be used to improve image-based modeling and rendering because they allow selection of pertinent and non redundant positions for the camera which will capture the images.

4.7 Total curvature-based global offline exploration

In this section is presented a technique for a offline global exploration of virtual worlds, based on the total virtual world curvature, as defined in chapter 3.

For this exploration, some aesthetic criteria, insuring the exploration quality, must be defined [8, 9, 10]:

1. The movie should not be very long, but it must show as much information as possible.
2. The operator (or an algorithm in our case, as the movie is to be created automatically) should avoid fast returns to already visited points.
3. The camera path must be as smooth as possible, because a film with brusque direction changes is confusing for the user.
4. The operator should try to guide the camera via viewpoints as good as possible.

4.7.1 Offline construction of a visibility graph

The exploration method is based on a *visibility graph* structure, defined as follows:

A visibility graph is a graph $G=(S \cup V, E)$, where:

S is the set of viewpoints on the discrete sphere where moves the camera,
V is the set of all vertices of the virtual world,
E is the set of arcs joining elements of S and V and representing visibility between elements from S and V.

This graph may be computed in $O(n_f \bullet \sqrt{n_s} \bullet n_v)$ operations in the worst case, where:
n$_f$ is the number of faces (polygons) of the virtual world,
n$_s$ is the number of viewpoints on the discrete sphere,
n$_v$ is the number of vertices of the virtual world.

The visibility graph is computed offline once for each virtual world to explore and used each time an exploration of the virtual world is needed.

4.7.2 Incremental computation of the camera trajectory

The method uses the *visibility graph* structure and a minimal set of viewpoints allowing to see all the vertices of the virtual world is computed in incremental manner. The main idea is to push the camera towards unexplored areas.

Thus, having a trajectory from the starting point to the current camera position, the camera is pushed towards pertinent viewpoints allowing to see the maximum of not yet seen details of the scene. To do this, at each step, a mass is assigned to each point of the discrete sphere and to the current position of the camera. The value of the mass assigned to a viewpoint is chosen according to the new information brought by the viewpoint. The camera position is then submitted to the Newton's law of gravity. The superposition of gravitational forces for the camera current position is the vector of movement.

More precisely, the total curvature-based global virtual world exploration technique works as follows:

- The main idea is to determine where lie unexplored areas of a scene, then to create "magnets" in these areas.

- The camera is to be considered as a "magnet" of opposite polarity. Magnetic forces of large unexplored areas will attract the camera. In order to simplify the computations gravitational forces are used instead of magnetic ones.

- The method is incremental, thus, having a trajectory line from the starting point to the current position of a camera, the camera is to be pushed towards unexplored areas. The aesthetic criteria could be satisfied with the following schema of exploration: at each step a mass is assigned to each point of the discrete sphere, surrounding the virtual world, and to the current position of the camera. Then the camera position is put under the Newton's law of gravity. The superposition of the gravitational forces for the camera point is used as the vector of movement.

- In order to estimate the mass assigned to each point of the discrete sphere, the following criterion is used: More a viewpoint brings new information, greater is the mass assigned to it.

With this method, the camera tries to move as quickly as possible to large uncovered areas, and does not return to visited areas, because the mass $m(p)$ of a visited point p is zero. Moreover, usually, the camera does not have any brusque trajectory changes except in the case, where the camera finds a very interesting object suddenly. For example, as soon as the camera sees the interior of a building through a window, a big attracting area disappears. So, the camera moves to see another uncovered area. Fig. 4.36(a) shows an example of this kind of problem.

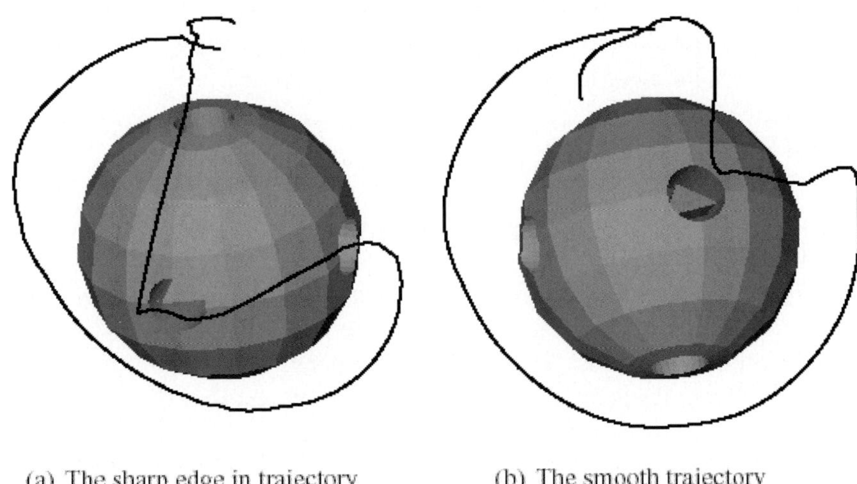

(a) The sharp edge in trajectory   (b) The smooth trajectory

Fig. 4.36: Trajectory with brusque change (a) and smoother trajectory using an inertia factor (Images by Dmitry Sokolov)

In order to attenuate the problem of brusque changes of the camera trajectory, an inertia factor is introduced. With this inertia factor, the camera trajectory becomes smoother, as it can be seen in Fig. 4.36(b).

On the other hand, the need of smooth camera movement is, maybe, not absolutely justified in the case of global virtual world exploration, because, as the camera always looks towards the center of the virtual world, brusque movement changes are not as disturbing as expected.

4.7.3 Further exploration of a virtual world

Sometimes it is not enough to view a virtual world only once. For example, if the user has seen some parts of the virtual world quickly, he (she), probably, did not understand it properly. So, it would be better to continue exploration of the virtual world.

The method presented above can be easily adapted to this task. If it is possible to "forget" parts already seen a long time ago, there will be always some regions of the object to explore.

The "forgetting" could be done in different manners. The simplest one is, probably, to determine the "size of memory" of the camera, that is, the time T, during which the algorithm keeps in mind a visited vertex.

4.7.4 Some results

This section presents some exploration trajectories computed for different virtual worlds.

Figure 4.37 shows a trajectory obtained with the "gravitational" total curvature-based method of global scene exploration. The images are snapshots taken consequently from the "movie". The first one corresponds to the start of the movement, i.e., to the best viewpoint. Let us compare it with the trajectory, shown in Fig. 4.9 and illustrating global incremental online exploration.

The trajectories are smooth and they have approximately the same length. However, it is easy to see that the new method gives a better path. Figure 3.9 shows that the camera passes under the floor of the virtual office, and during this time the user does not see the scene. The new trajectory is free of this disadvantage.

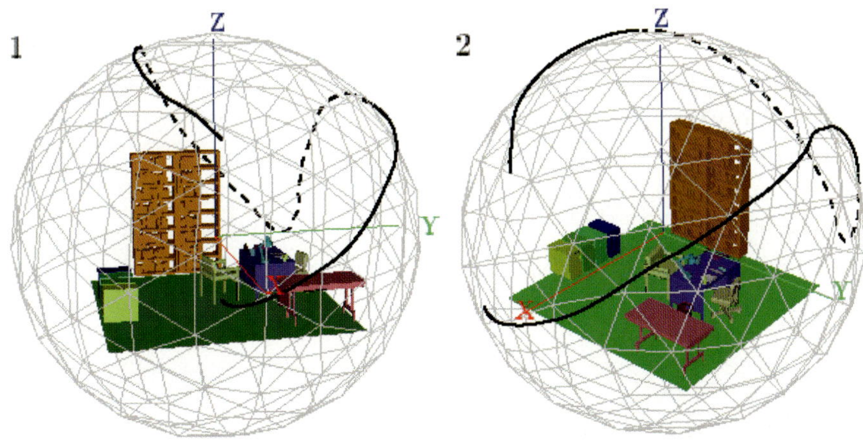

Fig. 4.37: Global offline exploration of a virtual office, using total curvature
(Images Dmitry Sokolov)

The next example of the new method application is shown at figure 4.38. The used model is very good for testing exploration techniques. It represents six objects inside a sphere with 6 holes.

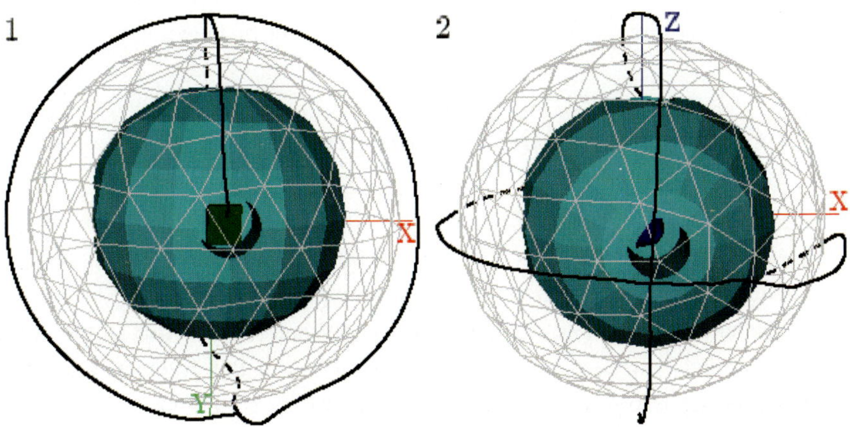

Fig. 4.38: Two views during exploration of the sphere model. All the objects inside the sphere are explored (Images Dmitry Sokolov)

Usual online global virtual world exploration techniques don't allow discovering all details of the scene, mainly because of missing information about unexplored areas. This offline method operates with the analytic visibility graph, created before camera path construction, which allows determining where unexplored areas are located. So, all details of the scene are seen during a minimal length exploration.

This result is not surprising, because offline virtual world exploration methods possess much more information than online methods, during exploration.

4.8   Local offline exploration

The local offline exploration method presented in this section uses a set of "photos", taken from previously chosen points of view, as a control set and creates walk-throughs simulating human paths. The camera is initially restricted to be at a constant height from the floor, in order to simulate the movement of a walking human.

The method is designed for environments with flat (or almost flat) terrains. In case when the ground is strongly deformed, the scene could be decomposed. In order to reduce the amount of computations, the set of viewpoints is sampled [9, 10].

4.8.1 Construction of the visibility graph

The construction of visibility graph, described in section 4.7 and in [8], can be easily adapted to calculate the visibility graph for the inside of a virtual world. In the adaptation of the algorithm the set of viewpoints is considered as a display to draw scene polygons. The algorithm iterates through each vertex of the scene and finds all pixels of the plane which are visible from the given vertex. Fig. 4.39 gives an illustration of the visibility graph construction method.

In local virtual world exploration it is not possible to keep only one direction of view, like in global exploration. So, a fixed set of 65 possible view directions is defined and used for each viewpoint. The possible view directions for the camera are shown in Fig. 4.40.

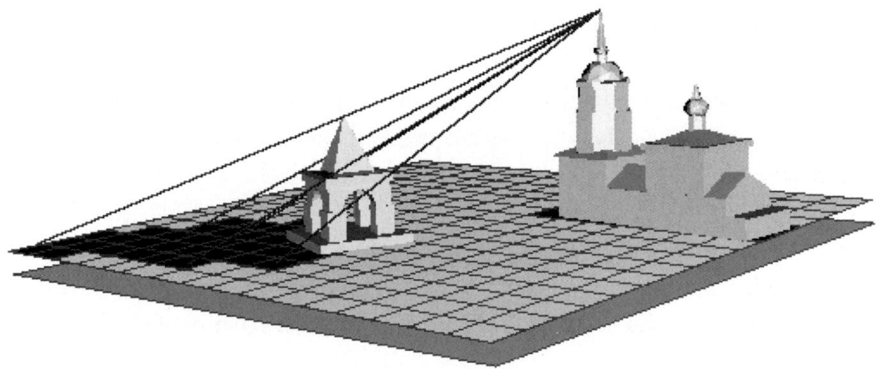

Fig. 4.39: Construction of the visibility graph. The church spire is visible from gray pixels and is not visible from the black ones (Image by Dmitry Sokolov)

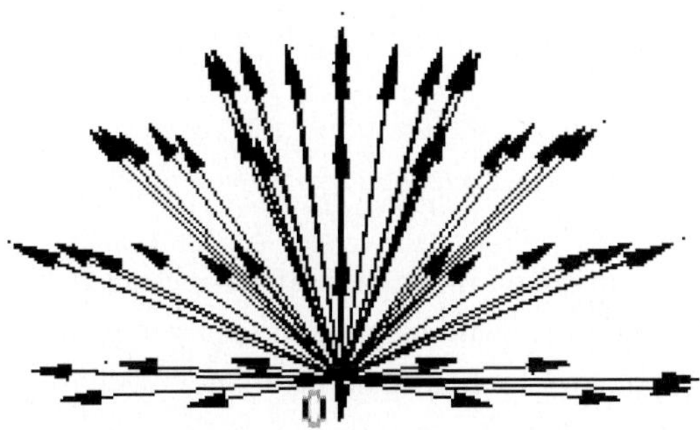

Fig. 4.40: 65 possible view directions for each point of view

4.8.2 Computation of pertinent views

The first step of the method is a search of a small number of pertinent "photos" (or views), that is viewpoints, on the plane of viewpoints, associated with view directions. To do this, a greedy algorithm is used: The best view is chosen at each step of the algorithm. Starting from the best viewpoint, the best viewpoint for the unexplored part of the virtual world is found at each step. The algorithm stops when the quality of the set becomes greater than a threshold value.

The used view quality criterion is the one defined in section 3.9 of chapter 3, based on object understanding. This step is the offline part of the method.

4.8.3 Computation of camera trajectory

The next step consists in computing of camera trajectory. The basic idea is to start with the viewpoints calculated during the first step, and to join the viewpoints with the camera path.

In order to avoid collisions, a visibility graph $G = (S, E)$ is calculated, where S is a set of viewpoints and E a set of arcs. An arc $(s1, s2)$ between two viewpoints $s1$ and $s2$ belongs to E if and only if there are not obstacles between $s1$ and $s2$.

As, for each viewpoint s, 65 view directions are associated, for a couple of viewpoints $(s1, s2)$ we can have 130 views. In order to select the best for the arc $(s1, s2)$, some additional costs and discounts are introduced:

3. The cost, proportional to Euclidian distance, of moving the camera from one point to another one.
4. The cost, proportional to the rotation angle, of turning the camera.
5. The discount that forces the camera to move via good viewpoints.

In Fig. 4.41 one can see an illustration of these costs and discounts. In this figure, paths 1 and 3 have the same length and viewpoints allow to see equal information. However, path 1 may be considered better, because path 3 requires more efforts for camera turns. Path 2 is shorter than path 1 but path 1 may be considered better than path 2, because some viewpoints of path 2 are too close to the building to explore and can see only a small part of the walls.

Now, the camera trajectory can be computed. The problem is to connect the already computed views. The problem may be solved by converting it to a travelling salesman problem, where the control set of views is the set of cities to visit. The travelling costs between pairs of cities are to be determined as the length of the shortest paths between them.

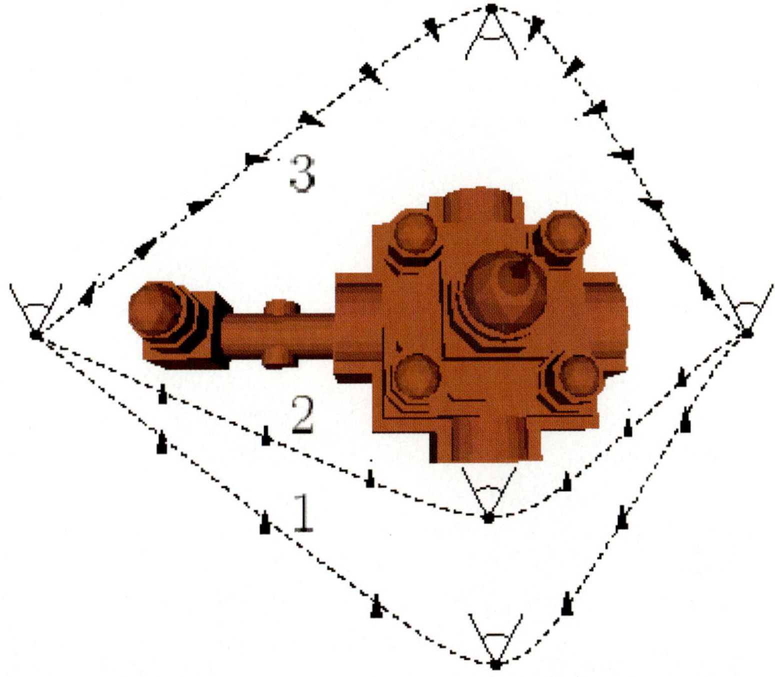

Fig. 4.41: Comparison of three exploration paths (Image by Dmitry Sokolov)

The final trajectory of the camera may be computed as the shortest Hamiltonian path. The only problem with this method is that the travelling salesman problem is NP-complete, even if there exist good approximation algorithms to resolve this problem.

The viewpoint visiting order is determined by the resolution of the travelling salesman problem. Some linking views may be added during the resolution process of the problem, in order to avoid collisions with obstacles.

Fig. 4.42 shows the final exploration trajectory for a virtual town. One can see the solid black triangles which were added during the problem resolution process. These triangles indicate additional viewpoints to visit in order to avoid collisions with objects.

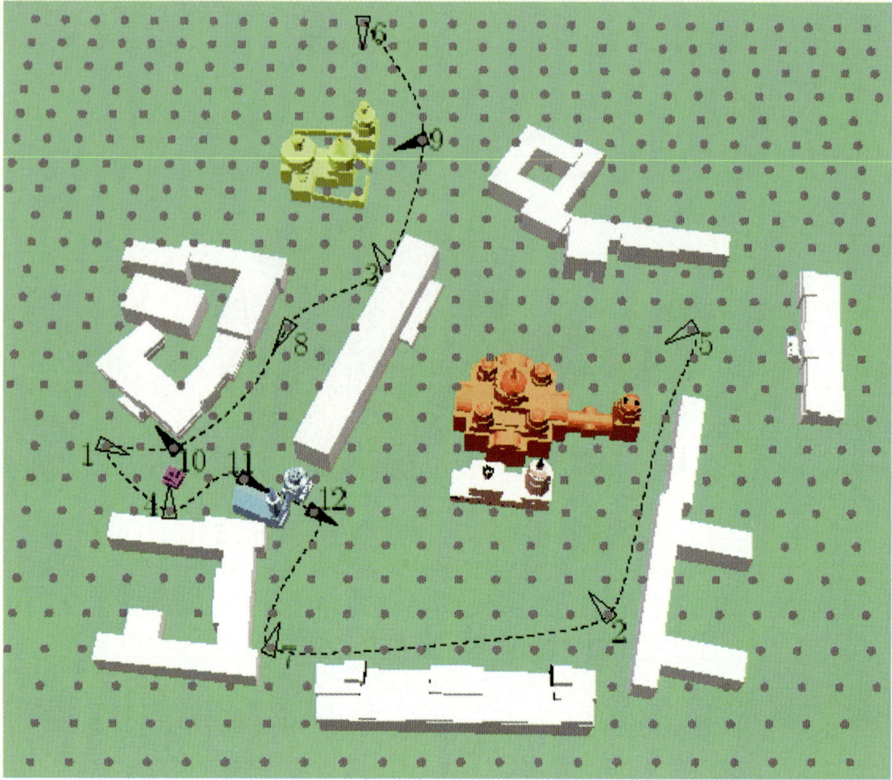

Fig. 4.42: Final camera trajectory for offline exploration of a virtual world. Black triangles correspond to new added views (Image by Dmitry Sokolov)

In the chosen solution it is sure that there are not collisions of the camera with obstacles. However, it is not always sure that brusque changes of direction are avoided. Such direction changes are disturbing for the user, especially during local virtual world exploration. A possible partial solution is to compute a spline, using the trajectory viewpoints as control points. This solution is not always satisfactory because collisions may appear, as the trajectory is changed.

4.8.4 Improvements

We have seen that, in the offline local exploration method described in this section, the camera can only move at a constant height from the floor. If we want to use all three dimensions for the camera, the viewpoint set has to be extended. The set of viewpoints has to fill all the 3D space. A possibility is to use a set of plane layers, where the viewpoints fill all the volume of the virtual world to explore. Fig. 4.43 shows this extended set of viewpoints.

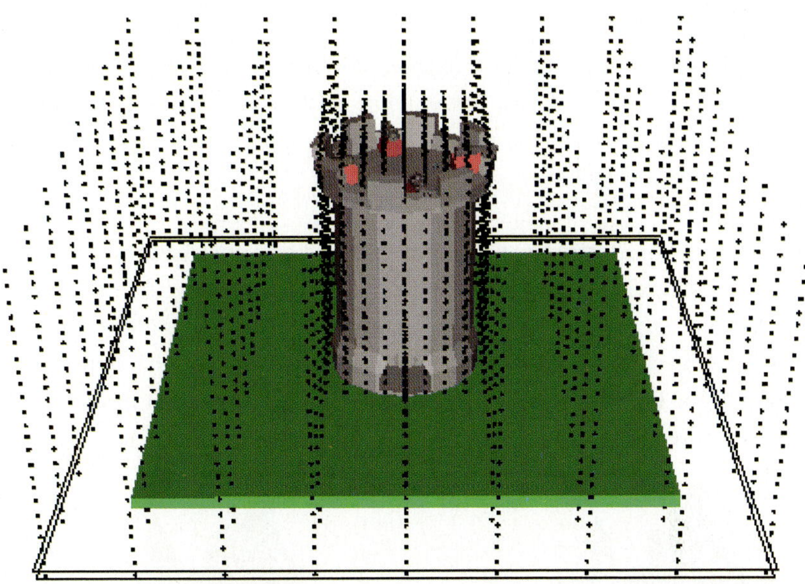

Fig. 4.43: Plane layers. Extension of the viewpoint set (Image by Dmitry Sokolov)

Fig. 4.44: Offline local exploration of a tower (Image by Dmitry Sokolov)

The method works in the same manner as the previous one. The only change is that the set of possible viewpoints is extended.

Fig. 4.44 shows an example of local exploration of a tower. The hidden part of the camera trajectory is shown in the white dashed line. In order to improve the presentation, are also drawn in white dashed line the treasure chest and the sword, which the camera is looking for.

Another proposed improvement of the method was introduction of the notion of "semantic distance", in order avoid some drawbacks in exploration. Indeed, during local exploration, the smoothness of camera path doesn't guarantees that exploration will not be confusing for the user. This is due to the fact that with this kind of exploration we have not only a list of points of view but a list of couples (viewpoint, view direction).

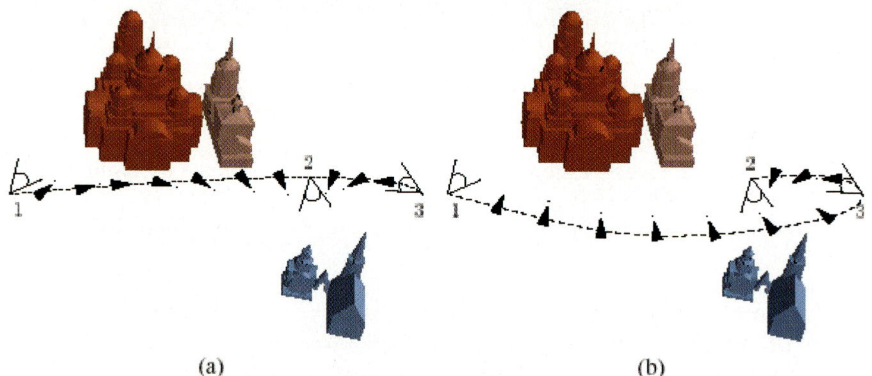

(a)   (b)

Fig. 4.45: Confusing and improved local exploration (Image by Dmitry Sokolov)

Figure 4.45(a) shows a virtual world comporting two churches. There are three manually fixed viewpoints. The dashed line is the trajectory, obtained by the above method of local exploration. The exploration is started at the viewpoint 1. It is easy to see that the camera turns aside from the big church in the viewpoint 2 and then reverts back again. The trajectory is very short and smooth, but the exploration is confusing for the user.

Fig. 4.45(b) shows another trajectory, where the exploration starts in the viewpoint 1, passes via the viewpoint 3 and terminates in the viewpoint 2. The camera turns aside from the big church only when its exploration is complete. Despite the increased length of the trajectory, the exploration is clearer.

Thus, it is possible to introduce a few semantics, in order to avoid this drawback. This is made by introducing the concept of *semantic distance*. The semantic distance between two concepts is proportional to the number of relationships between these objects. So, if two views show the same object, their distance is smaller than the distance between two views showing different objects.

4.9 Conclusion

In this chapter we have described several methods allowing exploration of virtual worlds by a moving camera, representing the user. The purpose of all these techniques is to allow a human user to visually understand the explored virtual world.

The common point of these methods is the use of the notion of viewpoint complexity and intelligent techniques, in order to determine interesting views for pertinent exploration of virtual worlds.

Two kinds of virtual world exploration have been presented. The first one, so-called *online exploration*, allows the user to immediately explore the chosen virtual world, whereas the second method, so-called *offline exploration*, uses a pre-processing step and prepares the virtual world for future explorations.

Some of described virtual world exploration methods allow *global exploration*, where the camera moves on the surface of a sphere surrounding the virtual world. The other methods allow *local exploration*, where the camera moves inside the virtual world.

We have chosen to describe methods, that we know very well, instead of making a state of the art of the virtual world exploration area. Of course, some other methods, allowing virtual world exploration, exist, but their purpose is often different from our, which is to allow a human user to well understand a virtual world.

Some other papers were published on the areas of camera positioning and camera path planing but their purpose is different from our, that is virtual visual world understanding for a human user. So, L. Fournier [18] gives a theoretical study of automatic smooth movement of a camera around a scene, based on reinforcement learning. J.-P. Mounier [19] uses genetic algorithms for planing paths for a camera walking inside a scene. In [16] a solver using interval arithmetic is used to obtain camera movements satisfying user-defined constraints. Some other papers [14, 15] try to include lighting in viewpoint quality criteria for virtual world exploration. This aspect will be seen in chapter 5.

In [17] a good State of the Art review on camera control techniques in Computer Graphics is presented. This revue is very general and its scope is much wider than automatic virtual world exploration.

References

1. Barral, P., Dorme, G., Plemenos, D.: Visual understanding of a scene by automatic movement of a camera. In: GraphiCon 1999, Moscow, Russia, August 26 - September 1 (1999)
2. Barral, P., Dorme, G., Plemenos, D.: Visual understanding of a scene by automatic movement of a camera. In: Eurographics 2000 (2000)
3. Barral, P., Dorme, G., Plemenos, D.: Intelligent scene exploration with a camera. In: International conference 3IA 2000, Limoges, France, March 3-4 (2000)
4. Dorme, G.: Study and implementation of understanding techniques for 3D scenes, Ph.D. thesis, Limoges, France, June 21 (2001)
5. Vázquez, P.P., Feixas, M., Sbert, M., Heidrich, W.: Viewpoint Selection Using Viewpoint Entropy. In: Vision, Modeling, and Visualization 2001, Stuttgart, Germany, pp. 273–280 (2001)
6. Vazquez, P.-P.: On the selection of good views and its application to computer graphics, PhD Thesis, Barcelona, Spain, May 26 (2003)
7. Vazquez, P.-P., Sbert, M.: Automatic indoor scene exploration. In: Proceedings of the International Conference 3IA 2003, Limoges, France, May 14-15 (2003)
8. Sokolov, D., Plemenos, D., Tamine, K.: Methods and data structures for virtual world exploration. The Visual Computer 22(7), 506–516 (2006)
9. Sokolov, D.: Online exploration of virtual worlds on the Internet, Ph.D thesis, Limoges, France, December 12 (2006)
10. Sokolov, D., Plemenos, D.: Virtual world exploration by using topological and semantic knowledge. The Visual Computer 24(3) (March 2008)
11. Jaubert, B., Tamine, K., Plemenos, D.: Techniques for off-line exploration using a virtual camera. In: International Conference 3IA 2006, Limoges, France, May 23–24 (2006)
12. Jaubert, B.: Tools for Computer Games. Ph.D thesis, Limoges, France (July 2008) (in French)

13. Plemenos, D., Benayada, M.: Intelligent display in scene modeling. In: International Conference GraphiCon 1996, St Petersburg, Russia (July 1995)
14. Lam, C., Plemenos, D.: Intelligent scene understanding using geometry and lighting. In: 10th International Conference 3IA 2007, Athens, Greece, May 30-31, pp. 97–108 (2007)
15. Vázquez, P.P., Sbert, M.: Perception-based illumination information measurement and light source placement. In: Proc. of ICCSA 2003. LNCS (2003)
16. Jardillier, F., Languenou, E.: Screen-Space Constraints for Camera Movements: the Virtual Camera man. Computer Graphics Forum 17(3) (1998)
17. Christie, M., Olivier, P., Normand, J.-M.: Camera Control in Computer Graphics. Computer Graphics Forum 27(8) (December 2008)
18. Fournier, L.: Automotive vehicle and environment modeling. Machine learning algorithms for automatic transmission control. PhD Thesis, Limoges, France (January 1996) (in French)
19. Mounier, J.-P.: The use of genetic algorithms for path planning in the field of declarative modeling. In: International Conference 3IA 1998, Limoges, France, April 28-29 (1998)
20. Desroche, S., Jolivet, V., Plemenos, D.: Towards plan-based automatic exploration of virtual worlds. In: International Conference WSCG 2007, Plzen, Czech Republic (February 2007)

5 Improvements and applications

Abstract. Visual (or viewpoint) complexity may be used in various computer graphics areas, in order to improve traditional methods. In this chapter, an improvement and three applications of viewpoint complexity will be presented. The proposed improvement ads lighting to viewpoint quality criteria. The three applications are image-based modeling, ray-tracing and molecular rendering, where the use of viewpoint complexity highly improves obtained results.

Keywords. Good point of view, Viewpoint complexity, Viewpoint entropy, Scene geometry, Lighting, Image-based modeling, Image-based rendering, Ray-tracing, Pixel super-sampling, Molecular rendering.

5.1 Introduction

In chapter 1 a definition, based on scene geometry, of visual (or viewpoint) complexity has been proposed. In chapter 3 and chapter 4, we have presented the main use of viewpoint complexity, scene understanding, more precisely static scene understanding and dynamic virtual world understanding, where visual intelligent exploration of the virtual world is required.

However, some questions remain open. The scene geometry-based viewpoint complexity (or quality) criteria are very important and allow efficient visual scene understanding. Meanwhile, taking into account only scene geometry is sometimes not full satisfactory. In real world, lighting is very important in understanding our environment and lack of lighting does not allow to see anything. So, in this chapter, we try to define a viewpoint quality criterion taking into account not only scene geometry but also lighting. The main advantage of the proposed method, which has to be improved, is that it integrates lighting in viewpoint quality computation, remaining compatible with the already seen best viewpoint estimation methods.

Traditional modeling methods in image-based modeling and rendering suffer from a big dilemma. If a great number of images is used in scene modeling, image-based rendering becomes time-consuming. On the other hand, if a small number of images (that is camera positions) are used, obtained by a uniform sampling

D. Plemenos and G. Miaoulis: Visual Complexity, SCI 200, pp. 155–177.

of the camera space, rendering may become very noisy, because of lack of information on several details of the scene. The use of viewpoint complexity allows to find a small set of pertinent camera positions, permitting to avoid, or to limit, noise in rendering.

In order to get high quality images with ray-tracing, pixel super-sampling is often required, because of aliasing problems inherent to ray-tracing. The problem is quite similar to that of image-based modeling and rendering, explained above. If pixel super-sampling is not applied, images may be noisy. On the other hand, if all the pixels of the image space are super-sampled, the rendering cost becomes exorbitant. The use of viewpoint complexity permits to choose pixels, or sub-pixels, which will be super-sampled and to apply adaptive pixel super-sampling.

Bio-informatics is a very important new research area, where the role of computer graphics is essential. Visualization of molecules, or of sets of molecules, allows a good understanding of their spatial structures. Viewpoint complexity may be used in this area in order to help choosing the appropriate camera position, depending on the current situation.

Other computing graphics areas may use viewpoint complexity and improve results obtained with traditional methods. The more important of these areas is doubtless Monte Carlo radiosity. Very important improvements may be obtained by using viewpoint complexity. Application of viewpoint complexity in Monte Carlo radiosity, due to its importance, will be described in chapter 6.

The chapter is organized in the following manner:

In section 5.2 the problem of integration of lighting parameters in viewpoint quality estimation is presented and a first approach to integrate light in viewpoint quality criteria is described. Then some other methods, more precise are proposed. In section 5.3 the problem of image-based modeling and rendering is presented and two methods improving performances of this rendering technique are proposed. In section 5.4 the problem of pixel super-sampling in ray-tracing is presented, as well as a method to reduce the super-sampling cost, while insuring image quality.

In section 5.5 application of viewpoint complexity in bio-informatics is described, where this criterion is used to improve rendering of molecules and proteins, with a different rendering mode, according to each case. Finally, we conclude in section 5.6.

5.2 Scene understanding and lighting

The problem of understanding a scene is currently a more and more pertinent problem, because of the development of web applications and possibilities, for a user, to discover new, never seen, scenes on the net, which are generally difficult to well understand without a tool able to evaluate the pertinence of a view, to choose a good view for each world and even to allow to explore it with a virtual camera.

With the fast development of computers capabilities this last decade, the problem of well understanding complex virtual worlds becomes more and more crucial and several recent papers try to take into account this problem.

In order to evaluate the pertinence of a view, current techniques consider that geometry and topology of a virtual world are important elements to take into account but the problem is how to do it. The problem is difficult to resolve because its solution consists to quantify the human perception of an image.

Even if the real human perception is not perfectly taken into account with current visual pertinence evaluation techniques, these techniques give generally results good enough allowing to apply these techniques in several areas, such as computer games, virtual museums visiting, molecules visualisation or realistic rendering.

However, it is easy to understand that the only knowledge of geometry and topology of a scene is not enough to allow precise quantification of the human perception. If the virtual world is illuminated, it is important to take into account illumination of its elements in evaluation of visual pertinence. Everybody knows that, even if there are lots of pertinent details in a dark room, no one of them is visible and it is not pertinent to choose a point of view allowing to look inside the dark room.

In this section we will present new techniques allowing to take into account the lighting parameters of a scene, in intelligent scene understanding and exploration, together with its geometry, in order to get more pertinent criteria for choosing viewpoints and exploration paths allowing to understand this scene. For the moment it is supposed here that the camera always remains outside the scene.

Starting from evaluation functions currently proposed to estimate the geometry-based quality of a point of view, illumination-based criteria are introduced and combined them in order to give the camera more realistic decision criteria during scene exploration. The first results obtained with these methods seem fully satisfactory and show that lighting parameters are well integrated in the viewpoint quality estimation process.

What is the lighting problem? There are rather two different problems, which have to be resolved in different manners. The first problem is *absolute light source placement*, that is, where to put a light source in the scene for a good understanding of it by a human user. The second problem is *taking into account light source position*, that is, for a scene lighted by a light source, where to put the camera in order to allow a good understanding for a human user.

5.2.1 Absolute light source placement

The problem is how to compute light source(s) position(s) in order to illuminate a scene in optimal manner. The resolution of this problem does not depend on the camera position. A good illumination of the scene should allow easier understanding by the user, if a camera explores the scene.

In the simple case of a single punctual light source, if only direct lighting is considered, the problem may be resolved in the same manner as the camera placement problem. What we have to do is to look for the best viewpoint from the light source.

In the general case, the problem is much more complex. Available today methods are not satisfactory. Most of them are based on inverse lighting techniques, where light source positions are deducted from the expected result. However, methods proposed by Poulingeas et al. [22] and Poulin et al. [23, 24] are not entirely satisfactory, especially because it is not easy to well describe and formalise the expected results.

Design Galleries [25] is a general system to compute parameters for computer graphics but computation is not fully automatic. Another not fully automatic system to compute light source positions is presented in [26]. The method presented in [32] is based on the notion of *light entropy* and automatically computes lighting parameters but results are not entirely satisfactory without the help of the user.

For a fixed point of view, the problem is to find an optimal position for the light source, in order to better understand the scene. For a punctual light source, if we have a function automatically computing the quality of a point of view by taking into account not only the geometry of the scene but also lighting, it is possible to compute the best position for the light source by using the heuristic search described in chapter 1, where the surrounding sphere of the scene is divided in 8 spherical triangles and the best one is subdivided in order to progressively reach a position with the best evaluation. This position will be the optimal light source position. In subsection 5.2.2, the problem of finding a viewpoint evaluation function allowing to take into account both geometry and lighting will be discussed.

5.2.2 Taking into account light source position

Up to now, in chapters 2, 3 and 4, we have considered that the quality of a viewpoint is based on the geometry of the scene to be seen. However, a scene is often illuminated and several details, considered important according to the scene geometry, may be not visible for a given position of the light source, because they are shadowed. It is clear that, in such a case, it is important to take into account lighting in the computation of the quality of view from a viewpoint. If the number of scene details seen from a point of view is important, lighting of each visible detail has to be taken into account.

The problem of taking into account light source placement is quite different from the absolute source placement problem. Here the purpose is to take into account light source position in order to compute more precisely the pertinence of a view. The question to answer is: *Given a viewpoint P and a light source position L, how to compute the pertinence of the view from this viewpoint?* The problem is difficult to resolve in the general case but solutions may be proposed for some simpler cases.

Thus, Vazquez et al. [16] have proposed a perception-based measure of the illumination information of a fixed view. This measure uses Information Theory concepts. The authors use, as unit of information, the relative area of each region whose colour is different from its surrounding.

5.2.2.1 A naive first approach

It is possible to propose a method to compute the pertinence of a given view, taking into account the position of one (or more) punctual light source for direct lighting [27, 15]. This method is inspired from the method of viewpoint evaluation used in [1] and [4]. The general idea of this method was proposed by Plemenos et al. in [14]. We have already seen that in the method proposed in [1] and [4], the following equation (1) is used to compute de viewpoint quality.

$$I(V) = \frac{\sum_{i=1}^{n} \left[\frac{P_i(V)}{P_i(V)+1} \right]}{n} + \frac{\sum_{i=1}^{n} P_i(V)}{r} \tag{1}$$

where: I(V) is the importance of the view point V,
         Pi(V) is the projected visible area of the polygon number i
         obtained from the point of view V,
         r is the total projected area,
         n is the total number of polygons of the scene.

In this formula, [a] denotes the smallest integer, greater than or equal to a, for any a.

In order to compute information needed by this equation, OpenGL and its integrated z-buffer is used as follows:

A distinct colour is given to each surface of the scene and the display of the scene using OpenGL allows to obtain a histogram (Fig. 5.1) which gives information on the number of displayed colours and the ratio of the image space occupied by each colour.

As each surface has a distinct colour, the number of displayed colours is the number of visible surfaces of the scene from the current position of the camera. The ratio of the image space occupied by a colour is the area of the projection of the viewpoint part of the corresponding surface. The sum of these ratios is the projected area of the visible part of the scene. With this technique, the two viewpoint complexity criteria are computed directly by means of an integrated fast display method.

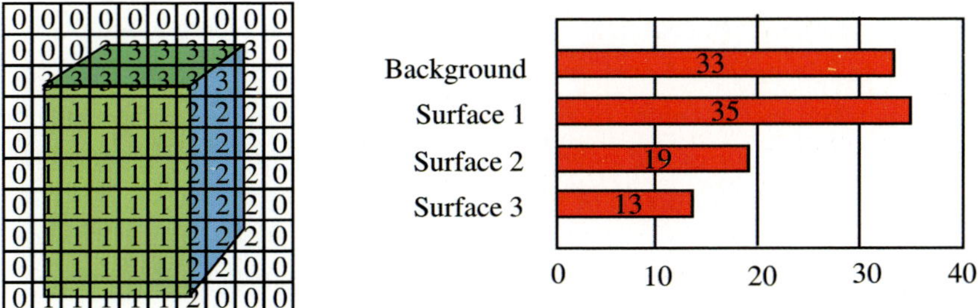

Fig. 5.1: Fast computation of number of visible surfaces and area of projected viewpoint part of the scene by image analysis.

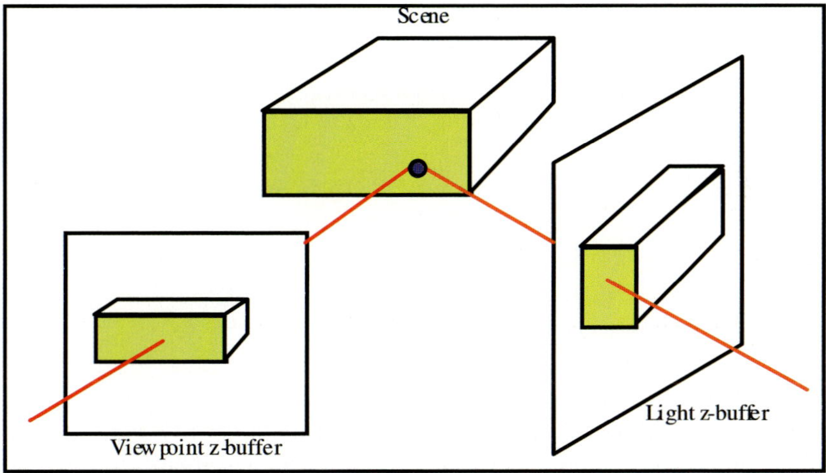

Fig. 5.2: Two z-buffers to estimate the quality of a viewpoint taking into account lighting of the scene.

Let us suppose that equation (1) is used to compute the quality of a point of view when only the geometry of the scene is used. In order to get an accurate estimation of the quality of view of a polygon of the scene from a given point of view it is important to integrate the quality of lighting of this polygon. A simple method to do this is to consider the angle of lighting from the light source to, say, the centre of the polygon and to introduce the cosine of this angle in equation (1).

In practice we can use two z-buffers, one from the point of view and one from the light source and approximate the cosine of the angle with the projected area of the polygon from the light source position (Fig. 5.2). For example, in equation (1), the considered visible projected area for a polygon will be de average value between the really visible projected area and the visible projected area if the light source position is taken as the centre of projection. That is, the number of pixels corresponding to the colour of the polygon of the first z-buffer will be added to the corresponding number of pixels of the second z-buffer and divided by 2.

In this method we use only the information that a polygon is lighted or not.

More generally, the formula to compute the quality of a point of view, taking into account the lighting, could be the following:

$$I_L(P, S) = \frac{\alpha\, I(P) + \beta\, I(S)}{\alpha + \beta}$$

where:
$I_L(P,S)$ is the global importance of a point of view P taking into account the light source S.
$I(X)$ is the geometry-based importance of the point of view X.
α and β are coefficients used to refine the respective contribution of geometry and illumination of the virtual world.

This method may be easily generalised for n punctual light sources.

5.2.2.2 Refining the method

The formula proposed in the previous subsection is a very general one. We are going to refine it in three manners:

1. The geometry-based importance of a point of view will be decomposed in number of visible polygons and projected visible area.
2. The relative importance of number of visible polygons and projected visible area will be made more precise.
3. The projected area of the visible part of the scene will be expressed in number of pixels, by explicitly using the OpenGL histogram, as explained above.

So, the final formula used to take into account lighting is the following:

$$I_L(P) = n\, p_e(P)\, c + n\, p_v(P)$$

Where:

$I_L(P)$ is the global importance of a point of view P taking into account the lighting.
n is the number of visible polygons of the scene from the viewpoint P.
$p_v(P)$ is the number of visible pixels from the viewpoint P.
$p_e(P)$ is the number of lighted pixels, visible from the viewpoint P.
c is a coefficient allowing to modify the importance of the lighting part of the formula. The best results are obtained with a value of about 1.9 for c.

In this formula, the light source position doesn't appear explicitly. However, it is implicitly expressed by the number of lighted pixels.

In this method, only direct lighting is taken into account.

5.2.2.3 Indirect lighting

In the methods presented in the previous subsections, only direct lighting is taken into account. Even if obtained results are satisfactory, sometimes, when a higher degree of precision is required, it is necessary to take into account indirect lighting. In such a case, the use of the OpenGL z-buffer is not sufficient. Other techniques, such as ray tracing or radiosity must be used. In this subsection a new good view criterion, taking into account both direct and indirect lighting,, is described, together with a method to evaluate the quality of a view.

The formula evaluating the view of a scene from a given viewpoint P becomes now:

$$I_L(P) = n\,(p_v(P) + c_1\,p_e(P) + c_2\,p_r(P) + c_3\,p_t(P))$$

Where:
$I_L(P)$ is the global importance of a point of view P taking into account direct and indirect lighting.
n is the number of visible polygons of the scene from the viewpoint P.
$p_v(P)$ is the total number of directly visible pixels from the viewpoint P.
$p_e(P)$ is the number of both directly and indirectly lighted pixels, visible from the viewpoint P.
$p_r(P)$ is the number of indirectly visible pixels from the viewpoint P by reflection.
$P_t(P)$ is the number of indirectly visible pixels from the viewpoint P by transparency.
c_1, c_2 and c_3 are coefficients allowing to modify the importance of indirect view and lighting. The best results are obtained with $c_1=2$, $c_2=1$ and $c_3=1$.

A ray tracing algorithm is used to evaluate the quality of a view. For each reflected or transmitted by transparency ray, if the ray intersects a polygon, $p_r(P)$ or $p_t(P)$ is incremented. If the indirectly visible polygon is lighted, $p_e(P)$ is also incremented.

5.2.2.4 Automatic computation of the best viewpoint

For a scene and a fixed punctual light source it is now easy to compute the best viewpoint, using one of the viewpoint quality criteria proposed in this subsection. The method proposed by Plemenos et al. [14, 1, 4] and described in chapter 3, may be used again. The scene is placed at the centre of a surrounding sphere whose surface contains all possible points of view. The sphere is initially divided in 8 spherical triangles and the best one, according to the used criterion, is selected and subdivided as long as the quality of obtained viewpoints is improved.

If neither light source nor viewpoint is fixed for the scene, the problem is to find on the surface of the surrounding sphere the couple (viewpoint, light source) which ensures the best evaluation, that is, the best view. There doesn't exist efficient manner to do this, even if we work on a discrete surface of the surrounding sphere. Every possible viewpoint has to be combined with all possible positions for the light source and the process is very time consuming.

5.2.2.5 Automatic scene exploration

For a scene and a fixed punctual light source, a variant of the method proposed in [1, 2, 3, 4] for incremental online exploration may be used. The best viewpoint position becomes the initial camera position and, for a given position of the camera, only a reduced number of next positions are considered, in order to warrantee smooth exploration, free of brusque changes of direction. Scene understanding may be improved by using plan-based incremental exploration proposed by S. Desroche et al. [21].

The problem with online exploration is that it supposes real time evaluation of a viewpoint. Exploration taking into account lighting is not yet implemented but it is sure that online exploration will not be possible if indirect lighting is taken into account, as presented in subsection 5.2.2.3.

5.2.3 First results

Methods and heuristics presented in subsection 5.2.2 have been implemented and obtained results allow to conclude that it is possible to integrate lighting parameters in estimation of the quality of a point of view.

In Fig. 5.3 to 5.6 one can see results obtained with the viewpoint quality criterion presented in subsection 5.2.2.2. In Fig. 5.3 the triceratops scene is seen from the computed best viewpoint with the best lighting, whereas in Figure 5.4 the same scene is seen from the same viewpoint with different positions of the light source. It is easy to see that the chosen viewpoint gives an interesting view of the scene.

In Fig. 5.5 the best view of the office scene is presented. In this scene the point of view is fixed and the viewpoint quality depends on the position of the light source. In Fig. 5.6, two other views of the office scene are presented with different positions of the light source. The point of view selected in Fig. 5.5 is not really the best one. What is presented is the best view from a chosen point of view, that is, the best position of the light source for this point of view. In Fig. 5.6 are presented other views from the same point of view but with different positions of the light source.

Fig. 5.3: Best view for the triceratops scene (Image C. Lam)

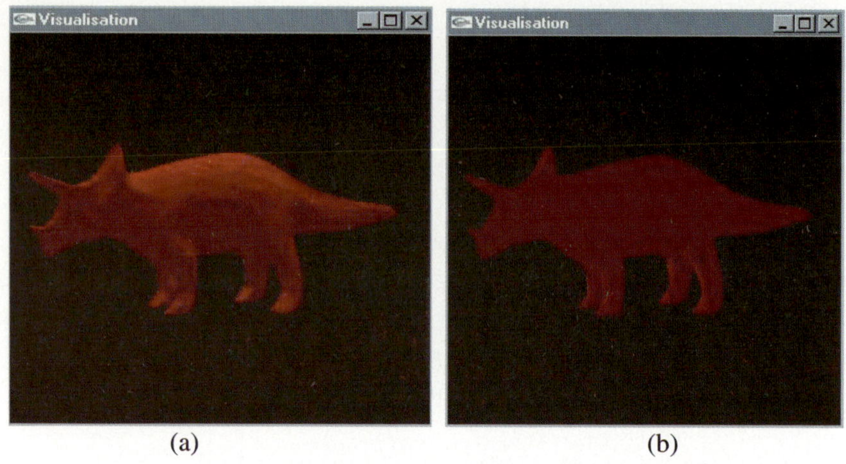

Fig. 5.4: Other views for the triceratops scene (Images C. Lam)

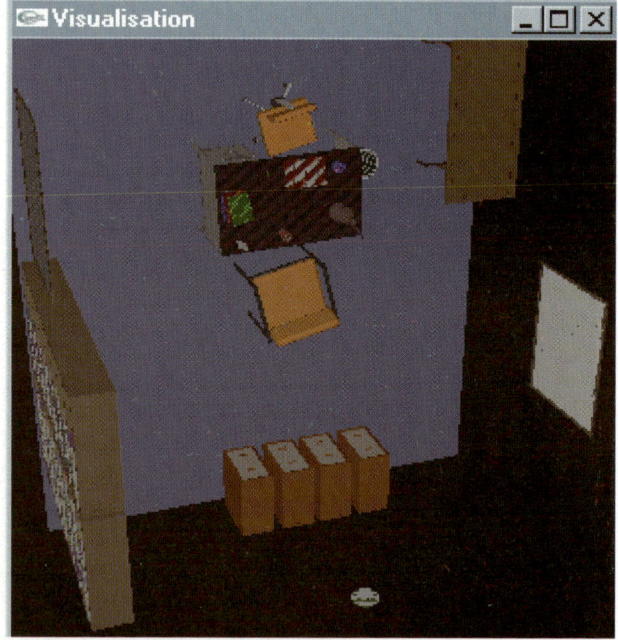

Fig. 5.5: Best view for the office scene (Image C. Lam)

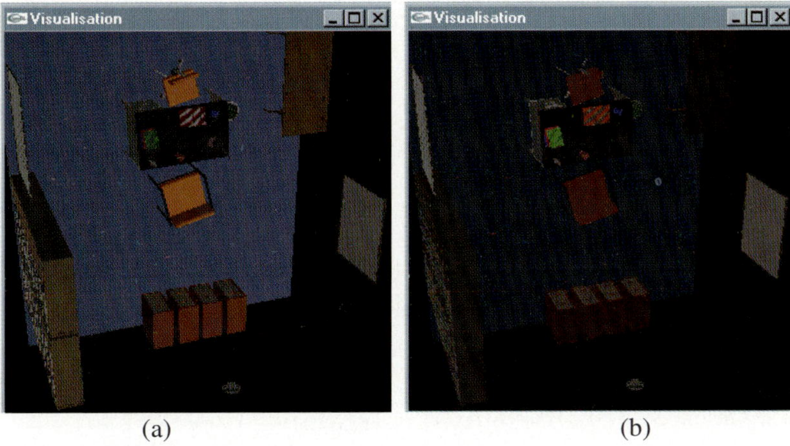

(a)   (b)

Fig. 5.6: Other views for the office scene (Images C. Lam)

The method of automatic computation of a good point of view based on the quality criterion of subsection 5.2.2.2 gives interesting and fast results, not dependent on the scene complexity by using the OpenGL z-buffer.

The main drawback of the method is that only direct lighting is taken into account. This limitation could imply less precision in results when considering scenes with many reflections and refractions.

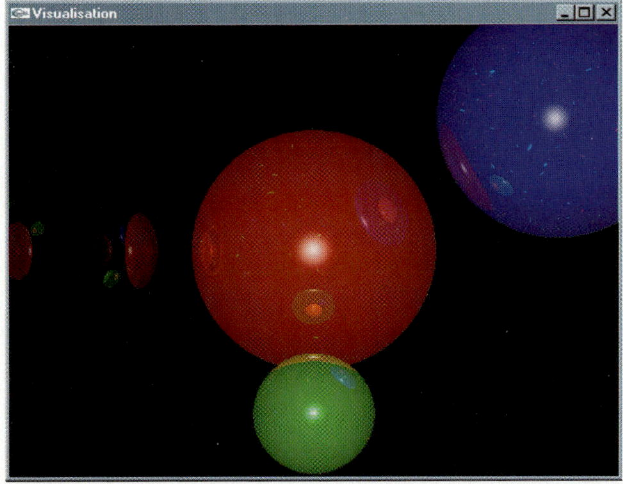

Fig. 5.7: Best view for a scene with reflections (Image C. Lam)

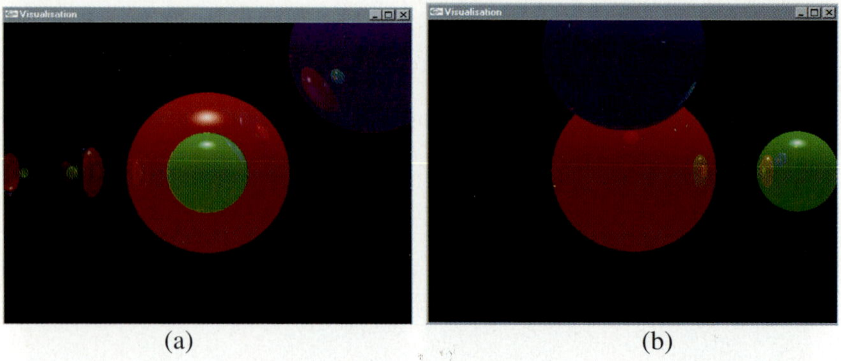

(a)   (b)

Fig. 5.8: Other views for a scene with reflections (Images C. Lam)

In Fig. 5.7, the best view is presented for the best viewpoint computation method based on the viewpoint quality criterion presented in subsection 5.2.2.3 and allowing to take into account direct and indirect lighting. In Fig. 5.8, two other, less interesting, views are presented. A ray-tracing algorithm is used to compute the evaluation function.

The method based on the viewpoint quality criterion integrating indirect lighting gives very interesting results but it is very time consuming, as it uses a ray tracing algorithm to compute reflections and refractions. For this reason it is difficult to use in online scene exploration.

5.2.4 Discussion

In this section, we have presented new methods permitting to take into account not only the geometry of a scene but also its lighting. Indeed, lighting is very important to make possible to a human user to understand a scene.

Two viewpoint quality criteria were proposed. The first one supposes that the scene is lighted by one (or many) punctual light source(s). Only direct lighting is considered. The method of computing a good viewpoint based on this criterion gives interesting results and is fast enough to be used in online scene exploration. The second viewpoint quality criterion supposes that the scene is lighted by one or more punctual light sources and takes into account both direct and indirect lighting. The method using this viewpoint quality criterion uses a ray-tracing algorithm in order to compute indirect lighting. This method allows to obtain very precise results but its drawback is that it is very time consuming, that is difficult to be used in online scene exploration. Of course, it is possible to use this method in off-line computation of a path for future exploration.

The described methods allow to mix geometric and lighting criteria in evaluation of a view. However, for the moment only punctual light sources are considered and taken into account by the proposed viewpoint quality criteria. It would be interesting to integrate non punctual light sources for more accurate results with realistic scenes.

5.3   Viewpoint complexity and image-based modeling

The purpose of image-based modeling and rendering is to replace the time consuming realistic rendering from a scene model, by a set of "photos", taken from various viewpoints.

This kind of modeling and rendering is very important in applications where real time rendering is requested, as for example in computer games, where continuous action requires real time display. Instead of a realistic image from each new position of the camera, the already stored ready to use images are displayed or, depending on the used viewpoint, a new image is quickly computed by interpolation between two existing images.

Image-based modeling and rendering are efficient techniques for real time rendering, only if the number of stored images is not too important, because, in such a case, the image retrieval process becomes very time consuming and it is not compatible with the real time constraint. So, the number of points of view has to be reduced but, on the other hand, the chosen viewpoints must cover all details of the scene.

Thus, in image-based modeling, it is important to compute a minimum set of points of view in order to use them in obtaining the set of images, which will replace the scene display.

5.3.1 Computing a minimal set of viewpoints – First method

How to find this minimal set of viewpoints? A first method to do this, based on viewpoint complexity, is the following:

1. Compute a sufficient number of points of view, according to the viewpoint complexity of the scene from a region of the sphere of points of view. A region is represented by a spherical triangle. The viewpoint complexity of a scene from a region may be defined as the average value of viewpoint complexities from the three vertices of the spherical triangle.

   - If the scene is viewpoint complex from the region, subdivide the region in 4 sub-regions.
   - Otherwise, add the center of the region to the current list of point of view.

2. Using an evaluation function for a set of points of view, try to eliminate elements of the list of points of view in several steps, by considering each point of view as a candidate to be eliminated. A point of view is eliminated if its contribution is empty (or quasi-empty) to the result of evaluation of the remaining point of view.

3. If, at the end of a step, no one point of view is eliminated, the process is finished and the current list of point of view contains a minimum set of points of view.

5.3.2 Computing a minimal set of viewpoints – Second method

A second method, also based on viewpoint complexity, is derived from an offline global virtual world exploration technique described in chapter 4. The method is resumed in the following lines [12, 13].

Computation of a minimal set of camera positions is performed in 4 steps:

1. Compute an initial set of camera positions by applying uniform sampling on the surface of a sphere surrounding the scene (Fig. 5.9).

2. Using an evaluation function, evaluate each camera position of the initial set and sort the set in decreasing order.

3. Starting from the sorted initial camera position set, create a new set by suppressing redundant camera positions, that is, positions allowing to see only details visible from the already selected positions in the set.

4. If the number of camera positions in the final set seems too important, apply
   once again step 3 to this set in order to get a really minimal set.

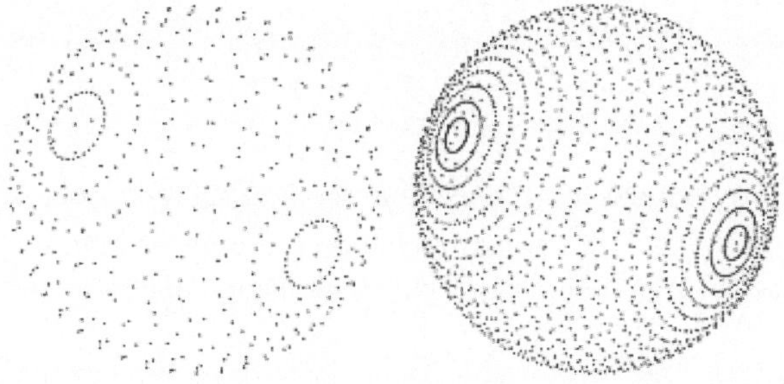

Fig. 5.9: Uniform sampling of the sphere surrounding the scene:

5.3.2.1 Evaluating camera positions

Each camera position is evaluated by means previously seen methods, especially
in chapters 2 and 3, based on viewpoint complexity.

Hardware-based fast viewpoint complexity evaluation may be used.

5.3.2.2 Computing a minimal set of camera positions

Starting from the sorted initial set a new reduced set of camera positions is created:

The current position in the initial set is compared to all the positions of the reduced set of camera positions. If all the visible details (surfaces and objects) from
the current position are visible from the camera positions already stored in the reduced set, the current position is ignored. Otherwise, it is stored in the reduced set
of camera positions.

The obtained reduced set is generally not minimal, so an additional processing
is required:

Every position of the reduced set is compared to all the other positions of the set and, if it doesn't allow to see more details than the other positions of the set, it is suppressed.

5.3.3 Some results

In Fig. 5.10 we see the reconstructed images using a classic approach (top) and a best viewpoint approach (bottom) with entropy as the measure of viewpoint complexity [8].

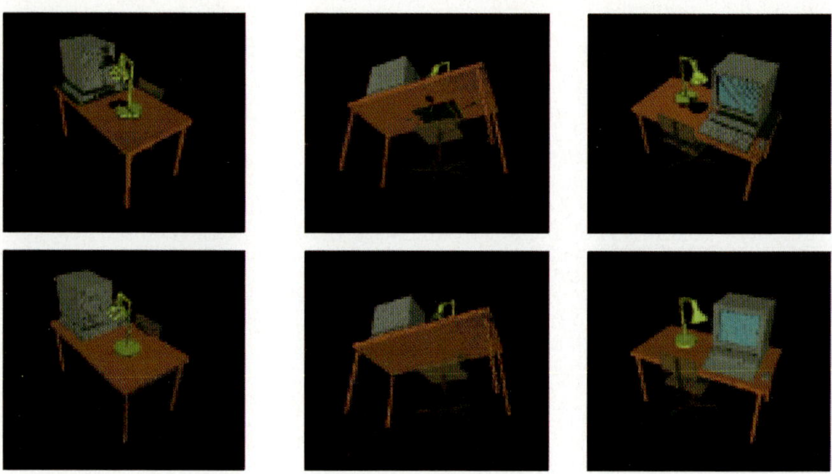

Fig. 5.10: Classic vs. best view entropy based capture for IBR representation (Images P.P. Vazquez)

It is easy to see that quality of images obtained with viewpoint complexity approach is much better than with the classic approach, where some parts of the image are missing because of lack of useful details of the scene.

5.4 Viewpoint complexity and ray-tracing

Obtaining a good quality image with ray-tracing demands to cast a lot of rays through each pixel of the screen plane. However, not all pixels need this amount of super-sampling. An homogeneous region will need less rays than a region with geometrical discontinuities and/or high illumination gradients. Viewpoint complexity, when restricted to a square area of pixels, to a pixel or sub-pixel area, can give a measure of the additional sampling necessity, and thus it can be the base for adaptive sampling methods.

Fig. 5.11: Uniform sampling (left) vs. Entropy based adaptive sampling (right) (images from the University of Girona)

Adaptive sampling methods for ray-tracing have been explored in [Ple87], where adaptive sampling was used in order to accelerate the ray-tracing process, by tracing less rays in regions where the scene is geometrically simple. Adaptive super-sampling has been explored in [29, 30] where entropy is used as the viewpoint complexity measure. In Fig. 5.11, on the right, we can see the improvement vs. uniform sampling, shown in Fig.5.11, on the left. The two images are very close to each other, but the uniform sampling-based method is much more time consuming than the viewpoint complexity-based adaptive sampling one.

5.5 Viewpoint complexity and molecular visualization

Visualization of molecules is relevant for molecular science, a discipline which falls in several areas such as Crystallography, Chemistry and Biology. Two kinds of views are important for scientists [31]:

1. Low entropy views (that is, low viewpoint complexity views) for a set of molecules (Fig. 5.12, left).
2. High entropy views (high viewpoint complexity views) for a single molecule (Fig. 5.12, right).

In the first case the views allow to see how the molecules arranges in space and thus infer physical properties.

The second case shows how the atoms are arranged in a molecule and allows to infer its chemical properties.

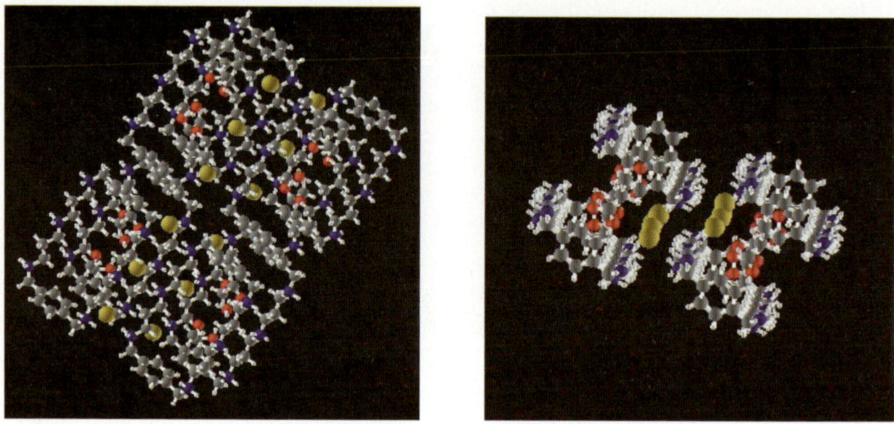

Fig. 5.12: Maximum (for a molecule) and minimum (for a set of molecules) viewpoint complexity views (Images P. P. Vazquez)

In Fig. 5.13 we see two elements of Carbon: graphite and diamond. From these views molecular scientists can infer the resistance to physical pressure. While diamond is very strong in three directions, the layered structure of graphite makes it easily exfoliable.

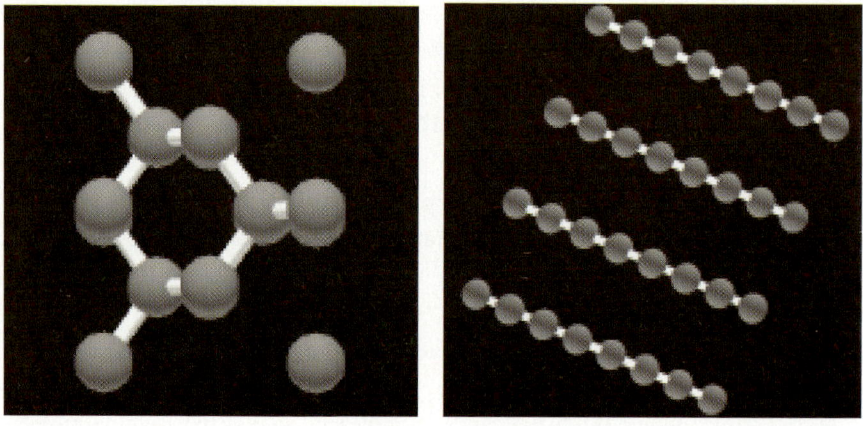

Fig. 5.13: Minimum entropy views for graphite and diamond (Images P. P. Vazquez)

5.6 Conclusion

In this chapter we tried to show the importance of viewpoint complexity, first by extending this concept and, second, by presenting three more areas where the use of viewpoint complexity improves results obtained by traditional computer graphics methods.

The extension of the concept of viewpoint complexity is based on introduction of lighting among the main criteria defining the quality of a view. Interesting results have been obtained with this extension but we must consider that it is possible to improve them, as the chosen values of coefficients in the used formulas are experimental ones and have to be refined. Virtual world exploration has also to be implemented, in order to visually verify the accuracy of used formulas and coefficients in such a context. Last but not least, it would be interesting to find improvements permitting to reduce the computation time when both direct and indirect lighting are used.

But viewpoint complexity may also be applied to other computer graphics areas, not only in scene understanding. Three application areas have been presented in this chapter.

The first application is a well known computing graphics area, image-based modeling and rendering. The main idea of the use presented in this chapter is to use viewpoint complexity in order to improve the scene covering and the number of stored images, by choosing a minimal set of good viewpoints, covering the scene.

Another application where viewpoint quality improves traditional techniques is ray-tracing. The method we have described uses viewpoint complexity to select some viewpoint complex regions of the scene. Adaptive pixel super-sampling is then performed in these regions, in order to improve the quality of the obtained image, without increasing too much the rendering cost.

The last presented application is molecular rendering. High viewpoint complexity camera positions are used to render (and understand) a single molecule, whereas low viewpoint complexity camera positions are used to render (and understand) a set of molecules.

A very important application of viewpoint complexity is intelligent Monte Carlo radiosity computation. We decided to dedicate a full chapter, chapter 6, to this viewpoint complexity application.

References

1. Barral, P., Dorme, G., Plemenos, D.: Visual understanding of a scene by automatic movement of a camera. In: GraphiCon 1999, Moscow, Russia, August 26 - September 1 (1999)
2. Barral, P., Dorme, G., Plemenos, D.: Visual understanding of a scene by automatic movement of a camera. In: Eurographics 2000 (2000)
3. Barral, P., Dorme, G., Plemenos, D.: Intelligent scene exploration with a camera. In: International conference 3IA 2000, Limoges, France, March 3-4 (2000)
4. Dorme, G.: Study and implementation of understanding techniques for 3D scenes, Ph.D. thesis, Limoges, France, June 21 (2001)
5. Vázquez, P.P., Feixas, M., Sbert, M., Heidrich, W.: Viewpoint Selection Using Viewpoint Entropy. In: Vision, Modeling, and Visualization 2001, Stuttgart, Germany, pp. 273–280 (2001)
6. Vazquez, P.-P.: On the selection of good views and its application to computer graphics, PhD Thesis, Barcelona, Spain, May 26 (2003)
7. Vazquez, P.-P., Sbert, M.: Automatic indoor scene exploration. In: Proceedings of the International Conference 3IA 2003, Limoges, France, May 14-15 (2003)
8. Vázquez, P.P., Feixas, M., Sbert, M., Heidrich, W.: Automatic View Selection Using Viewpoint Entropy and its Application to Image-Based Modeling. Computer Graphics Forum (December 2003)
9. Sokolov, D., Plemenos, D., Tamine, K.: Methods and data structures for virtual world exploration. The Visual Computer 22(7), 506–516 (2006)
10. Sokolov, D.: Online exploration of virtual worlds on the Internet, Ph.D thesis, Limoges, France, December 12 (2006)
11. Sokolov, D., Plemenos, D.: Virtual world exploration by using topological and semantic knowledge. The Visual Computer 24(3) (2008)
12. Jaubert, B., Tamine, K., Plemenos, D.: Techniques for off-line exploration using a virtual camera. In: International Conference 3IA 2006, Limoges, France, May 23 – 24 (2006)
13. Jaubert, B.: Tools for Computer Games. Ph.D thesis, Limoges, France (July 2008) (in French)
14. Plemenos, D., Benayada, M.: Intelligent display in scene modeling. In: International Conference GraphiCon 1996, St Petersburg, Russia (July 1995)
15. Lam, C., Plemenos, D.: Intelligent scene understanding using geometry and lighting. In: 10th International Conference 3IA 2007, Athens, Greece, May 30-31, 2007, pp. 97–108 (2007)
16. Vázquez, P.P., Sbert, M.: Perception-based illumination information measurement and light source placement. In: Proc. of ICCSA 2003. LNCS (2003)
17. Jardillier, F., Languenou, E.: Screen-Space Constraints for Camera Movements: the Virtual Cameraman. Computer Graphics Forum 17(3) (1998)

18. Christie, M., Olivier, P., Normand, J.-M.: Camera Control in Computer Graphics. Computer Graphics Forum 27(8) (December 2008)

19. Fournier, L.: Automotive vehicle and environment modeling. Machine learning algorithms for automatic transmission control. PhD Thesis, Limoges, France (January 1996) (in French)

20. Mounier, J.-P.: The use of genetic algorithms for path planning in the field of declarative modeling. In: International Conference 3IA 1998, Limoges, France, April 28-29 (1998)

21. Desroche, S., Jolivet, V., Plemenos, D.: Towards plan-based automatic exploration of virtual worlds. In: International Conference WSCG 2007, Plzen, Czech Republic (February 2007)

22. Jolivet, V., Plemenos, D., Poulingeas, P.: Inverse direct lighting with a Monte Carlo method and declarative modelling. In: Sloot, P.M.A., Tan, C.J.K., Dongarra, J., Hoekstra, A.G. (eds.) ICCS-ComputSci 2002. LNCS, vol. 2330, p. 3. Springer, Heidelberg (2002)

23. Poulin, P., Fournier, A.: Lights from highlights and shadows. Computer Graphics 25, 2 (1992)

24. Poulin, P., Ratib, K., Jacques, M.: Sketching shadows and highlights to position lights. Computer Graphics International 97 (June 1997)

25. Marks, J., Andalman, B., Beardsley, P.A., Freeman, W., Gibson, S., Hodgins, J., Kang, T., Mirtich, B., Pfister, H., Ruml, W., Ryall, K., Seims, J., Shieber, S.: Design galleries: A general approach to setting parameters for computer graphics and animation. In: International Conference SIGGRAPH 1997 (1997)

26. Halle, M., Meng, J.: Lightkit: A lighting system for effective visualization. IEEE Visualization (2003)

27. Plemenos, D., Sokolov, D.: Intelligent scene display and exploration. In: STAR Report. International Conference GraphiCon 2006, Novosibirsk, Russia, June 30- July 5 (2006)

28. Plemenos, D.: Selective refinement techniques for realistic rendering of 3D scenes. International Journal of CAD and Computer Graphics 1(4) (1987) (in French)

29. Rigau, J., Feixas, M., Sbert, M.: New Contrast Measures for Pixel Supersampling. In: Advances in Modeling, Animation and Rendering. Proceedings of CGI 2002, Bradford, UK, pp. 439–451. Springer, London (2002)

30. Rigau, J., Feixas, M., Sbert, M.: Entropy-Based Adaptive Sampling. In: Graphics Interface 2003, Halifax, Canada (June 2003)

31. Vázquez, P.P., Feixas, M., Sbert, M., Llobet A.: Viewpoint Entropy: A New Tool for Obtaining Good Views for Molecules. In: VisSym 2002 (Eurographics - IEEE TCVG Symposium on Visualization), Barcelona, Spain (2002)

32. Gumhold, S.: Maximum entropy light source placement. In: IEEE Visualization, October 27 - November 1 (2002)

6 Viewpoint Complexity in Radiosity

Abstract. In this chapter we present importance-driven subdivision techniques for efficient computing of radiosity. The purpose of these techniques is to perform an intelligent selection of interesting directions from a patch, based on viewpoint complexity. The first technique uses the importance of a direction to approximately compute form factors by adaptive subdivision of a hemi-cube. The second technique uses a heuristic function, which estimates viewpoint complexity in a given direction from a patch and recursively subdivides a hemisphere. This technique permits to improve the classical Monte Carlo progressive refinement technique by choosing good directions to send rays. The third technique is an improved variant of the second one. The heuristic function used to estimate the visual complexity in now applied to a region delimited by a pyramid and the visibility criterion used by the function is more accurate and easier to compute. A fourth technique, based on more accurate estimation of viewpoint complexity, is also proposed. This technique seems interesting but it is not yet implemented.

Keywords. Radiosity, Visibility, Viewpoint complexity, Heuristic search, Progressive refinement, Monte-Carlo radiosity, Form factors.

6.1 Introduction

Radiosity has been introduced in computer graphics in 1984 by Goral et al. [5] borrowed from thermal engineering for realistic image synthesis of scenes with diffuse inter-reflections. For radiosity computation, the form factor between each pair of elements of the scene (patches) must be obtained. This implies computing visibility between each pair of patches and is responsible for about 80% of the cost of the algorithm. The results of research in this area have enhanced the applicability of radiosity. In this direction we may underline adaptive subdivision [1], progressive refinement [2] and hierarchical radiosity [6]. Nevertheless, visibility computation remains the important parameter in the cost of radiosity algorithms.

To the authors' knowledge, up to now, it has not been proposed any strategy which takes really advantage of the knowledge of the scene in order to either reduce the number of rays or optimize the order of tracing them for the computation of visibility (form factors), with the exception of the papers of Plemenos and

D. Plemenos and G. Miaoulis: Visual Complexity, SCI 200, pp. 179–220.
springerlink.com   © Springer-Verlag Berlin Heidelberg 2009

Pueyo [10]; Jolivet and Plemenos [17]; Jolivet, Plemenos and Sbert [20, 21]; Plemenos [22], where heuristic methods were proposed, based on the techniques used for automatically finding *good view directions* [7, 3, 9] and, especially, on viewpoint complexity. Some authors have understood the importance of this problem but the proposed solutions are not entirely satisfactory. Thus, two papers have introduced the concept of visibility complex for efficient computation of 2D and 3D form factors [13, 18]. Another paper [19] introduces the notion of importance-driven stochastic ray radiosity, an improvement of the stochastic ray radiosity. The purpose of this chapter is to present techniques which perform intelligent selection of interesting directions from a patch, in order to improve energy distribution in radiosity algorithms. The techniques presented in this paper permit to obtain a useful image sooner than with classical methods, with, in most cases, an acceptable memory cost.

The main part of this work is divided into 4 sections. In section 5.2, after a brief presentation of the basic concepts of radiosity and the hemi-cube method for computing form factors, we will present a method for approximately computing form factors by an importance-driven adaptive hemi-cube subdivision. In section 5.3 we will describe some well known, Monte Carlo based, radiosity computing techniques and then we will present a new technique, the hemisphere subdivision technique, which permits to perform intelligent selection of directions to which rays will be sent and improves the Monte Carlo progressive refinement algorithm. Two versions of the hemisphere subdivision technique are presented, the second version being an improved version of the first one. Section 5.4 describes another technique for computing radiosity, based on hemisphere subdivision but using pyramids to determine a region and to estimate the importance of this region.

6.2 Radiosity and form factors

The radiosity algorithm is a method for evaluating light intensity at discrete points of ideal diffuse surfaces in a closed environment [5]. This environment is composed of polygons which are split into patches. The relationship between the radiosity of a given patch and the remainder is given by the following formula:

$$B_i = E_i + P_i \sum_{j=1}^{n} F_{ij}\, B_j$$

where :

Bi = radiosity in surface i

Ei = emissivity of surface i

Pi = reflectivity of surface i

Fij = form factor between surface i and j

n = number of surfaces in the environment

A form factor is given by the geometric relationship between two patches and represents the ratio of energy leaving one and arriving to the other [5].

   The form factor between finite patches is defined by:

$$F_{A_i \bar{A}_j} = \frac{1}{A_i} \int_{A_i} \int_{A_j} \frac{\cos\alpha_i \, \cos\alpha_j \, HID}{\pi r^2} \, dA_j \, dA_i$$

Where the function HID represents the visibility between the patches Ai and Aj.

The progressive approach [2] avoids the enormous cost of radiosity by computing form factors on the fly. Now, instead of keeping the whole system of equations to complete the energy exchange of the environment, a useful solution is computed. This is obtained by only shooting the energy of a reduced number of patches, the ones which contribute the most to the illumination of the environment.

6.2.1   The Hemi-cube Method for Computing Form Factors

In order to compute the form factors between an emitter patch and the other patches of the environment, the idea was to put a hemisphere to the center of the emitter patch and to project all the other patches onto the surface of this hemisphere. So, the form factor between the emitter patch and each other patch of the scene was proportional to the projected area of the other patch on the hemisphere's basis (Nusselt's analogy).

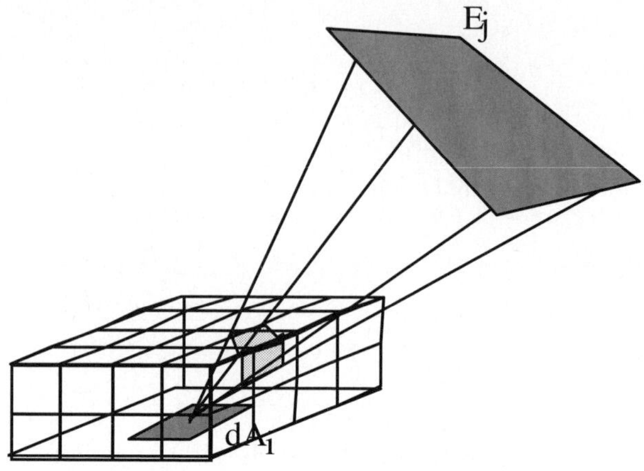

Fig. 6.1: Form factors calculation using the hemi-cube method

As the projected area of a patch on a hemisphere is rather difficult to compute, Cohen [16] had the idea to replace hemisphere by a hemi-cube. Each surface of the hemi-cube is divided into square cells. Only one surface, the visible one, is associated with each cell (Fig. 6.1).

6.2.2   The hemi-cube selective refinement technique

The hemi-cube technique allows easy computation of form factors but it requires projection of all the surfaces of the scene. This time consuming process could be useless in some cases. If a large visible surface is projected on the major part of a hemi-cube surface, its projection should normally avoid projection of several surfaces of the scene. So, the purpose of the method presented in this section is to reduce the number of projected surfaces by applying selective refinement techniques.

6.2.2.1 Selective Refinement

Selective refinement techniques were presented [14, 15] in order to improve previewing with ray tracing algorithms. These techniques used the notion of "macro-pixel", a square with a side of 2^n pixels. A ray is traced to each corner of every macro-pixel and, if visible surfaces are the same from the four corners, this surface is considered visible from all the pixels of the square. Otherwise, the macro-pixel is divided into 4 new macro-pixels and the same process is applied (Fig. 6.2).

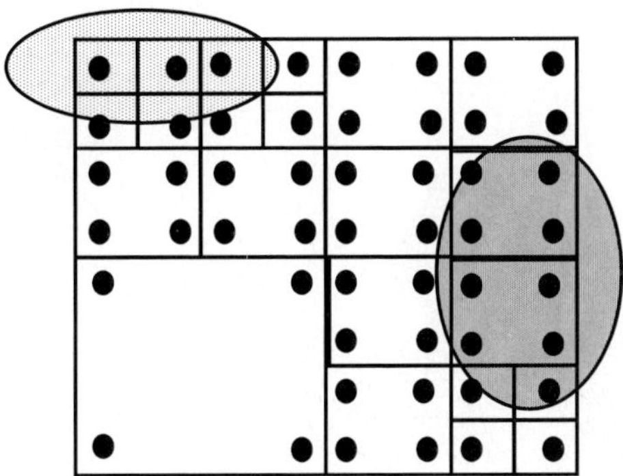

Fig. 6.2: Selective refinement for ray tracing

6.2.2.2 Selective Refinement for Computing Form Factors

We propose to apply selective refinement techniques to the computation of form factors by means of a hemi-cube.

The proposed method uses the surfaces of the hemi-cube instead of macro-pixels and works as follows:

1. Trace a ray from the centre of the emitter patch to each corner of each one of the 5 surfaces of the hemi-cube.
2. Determine the visible surface for each corner cell of the hemi-cube surface.
3. If the visible surface is the same for the 4 corner cells, there is a great probability for this surface to be visible from all the cells of the hemi-cube surface. Thus, the surface is associated with all the cells of the surface.
4. Otherwise, the surface is subdivided into 4 parts and the same process is restarted for each part (Fig. 6.3).

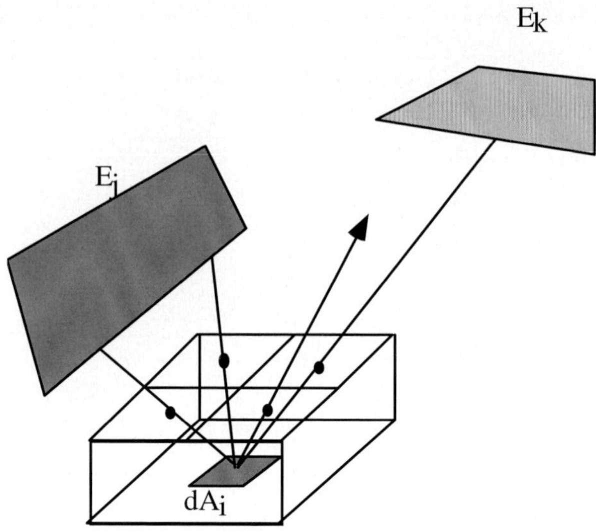

Fig. 6.3: The surface of the hemi-cube is divided into 4 parts

This method allows not to explicitly calculate visibility for each cell of the hemi-cube surfaces and thus to improve the speed of algorithms using the hemi-cube technique in form factors calculation. Moreover, it avoids, in several cases, useless projections of non visible surfaces. The obtained results are very interesting. Fig. 6.4 shows the gains obtained by the adaptive selective refinement technique compared to the classical hemi-cube technique for the scene of Fig. 6.6, while Fig. 6.5 gives a graphical comparison of the two techniques.

Number of steps	Classical hemi-cube	Adaptative hemi-cube
10	5 s	5 s
20	11 s	9 s
30	28 s	24 s
40	53 s	45 s
50	2 min 16 s	1 min 45 s

Fig. 6.4: Comparison of the two hemi-cube techniques

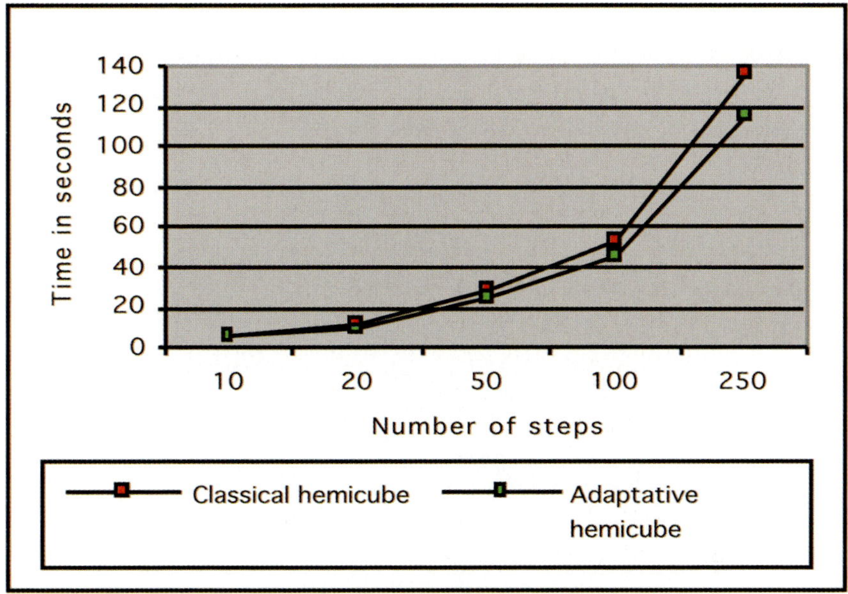

Fig. 6.5: Graphical comparison of the two methods

Fig. 6.6: The test scene (Image Vincent Jolivet)

Normally, the adaptive selective refinement technique could fail in some cases because it is possible that an object lies in the pyramid defined by the center of a patch and the four corner cells of a surface of the associated hemi-cube. Thus, if the visible surface is the same for the 4 corner cells (see Fig. 6.7), this small object can disappear. For this reason, this method should be used carefully. However, we have not encountered any problem with our test scenes, probably because there is important space overlapping among hemi-cube pyramids of neighboring patches. Objects missed by the hemi-cube of a patch will be taken into account by one or more hemi-cubes of neighboring patches (see Fig. 6.8).

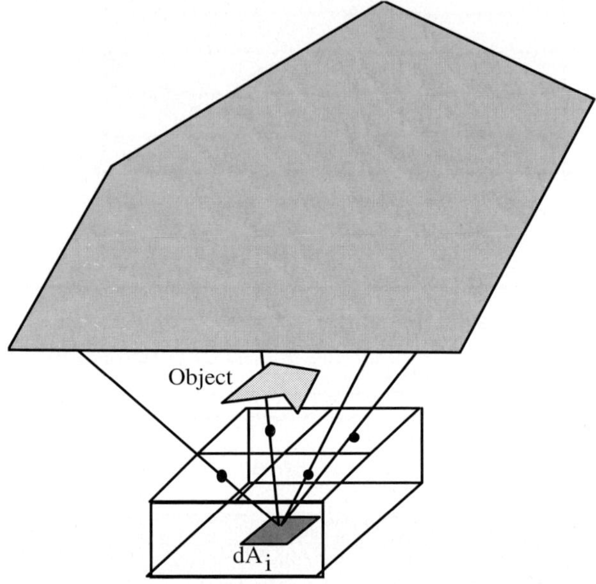

Fig. 6.7: A small object can disappear

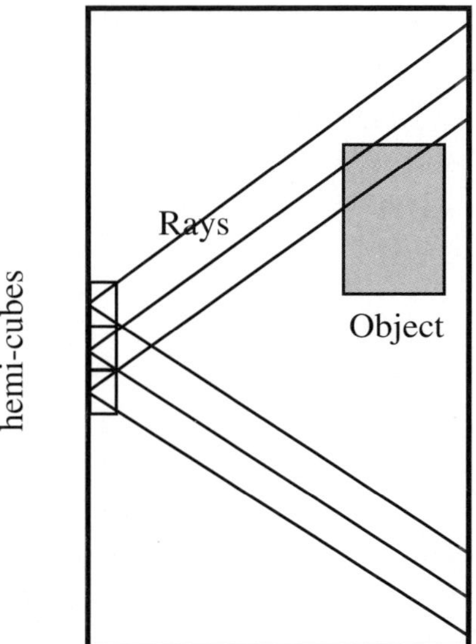

Fig. 6.8: 2D space overlapping among neighboring patches. The object is missed by the pyramid of the first hemi-cube but it is intersected by the pyramids of neighboring hemi-cubes.

Moreover, in order to increase the accuracy of this method, it is possible to subdivide each surface of the hemi-cube up to a chosen subdivision level before starting the selective refinement process.

6.3   Hemisphere subdivision techniques

6.3.1 Progressive refinement and Monte Carlo-based techniques

As stated in the previous section, radiosity is computed in environments composed of surfaces which are divided into patches. The purpose of radiosity improvement techniques is to obtain as fast as possible a useful image. Most of radiosity computation techniques are progressive techniques, where the radiosity of a scene is computed and refined in several steps.

6.3.1.1 Progressive Radiosity

The energy of each patch is distributed, starting from patches which contribute the most to the illumination of the environment [2]. The form factors are explicitly computed between the emitting patch and each one of the other patches of the environment.

An image of the scene is displayed each time a patch has been processed. So, the displayed image is progressively improved.

The radiosity computing process is achieved when all of the scene energy has been distributed. It is also possible to stop the radiosity process if a useful image is obtained.

6.3.1.2 Monte-Carlo Radiosity

Starting from the patch Ai which possesses the maximum of energy (Ai.Bi), a great number of rays is sent from randomly chosen points of each patch towards directions chosen according to a cosine distribution [11]. Each ray distributes the same amount of energy. The rays that are shot are supposed to distribute the whole energy of the patch.

Form factors are not explicitly computed. The fraction of the energy received by a patch is proportional to the number of received rays and to the energy diffused by each received ray.

An image of the scene is displayed each time the whole energy of a patch has been distributed. So, the displayed image is progressively improved.

The radiosity computing process is achieved when all of the scene energy has been distributed or when a useful image has been obtained.

6.3.1.3 Monte-Carlo Progressive Radiosity

An additional loop is added to the pure Monte-Carlo algorithm. At each step, the energy of a reduced number of additional rays is shot from each patch of the scene [4]. An image is displayed only at the end of a step, that is when the whole energy of a reduced number of rays has been totally distributed in the scene.

If the total number of rays shot until the current step (included) is k times greater than the number of rays shot until the previous step, the energy received by each patch until the previous step and the energy distributed by the current patch are divided by k and the remaining energy is then shot by the additional rays sent in the current step.

The whole process can be stopped if the produced image is considered useful.

6.3.2 The first hemisphere subdivision technique

The technique presented in this section is an adaptive technique permitting to improve the progressive refinement Monte-Carlo radiosity by optimising the choice of shooting rays. This approach, which was initially presented in [17], divides a spherical triangle using a heuristic based on the density of the scene in a given direction. The target of this technique is to reduce, at least in a first step, the number of rays shot from an emitter in order to obtain a useful image.

In the progressive refinement approach proposed in [4] for Monte Carlo radiosity, the algorithm starts distributing the energy using a reduced number of rays. Afterwards the number of rays is increased in order to refine the solution.

So it would be interesting to find "intelligent" algorithms which avoid tracing useless rays for visibility computation, in order to obtain a good image sooner.

The goal is to reduce the number of rays by performing a good choice of them. Thus, it is possible that in some directions a reduced number of rays is enough while in other directions more rays are necessary. The chosen strategy, based on a kind of viewpoint complexity, is to send more rays towards the directions where more surfaces will be found. This strategy is justified by the fact that the probability to meet a surface is greater for a ray sent to a direction where there are many surfaces than for a ray sent to a direction where the number of surfaces is small. Thus, let us consider the hemisphere surrounding the emitting patch (Fig. 6.9).

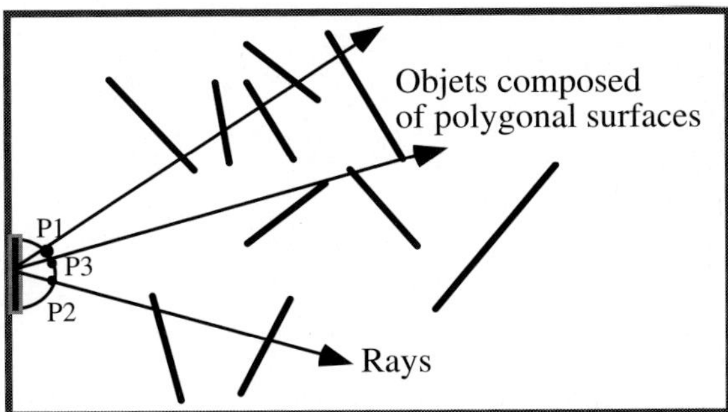

Fig. 6.9: Casting selected rays.

The following method is used to select new rays to shoot: if the number of surfaces (patches) intersected by a ray through point P1 in the hemisphere is ns(P1) and the number of surfaces intersected in the direction defined by point P2 is ns(P2); then a new ray will be traced into the direction defined by P3 where P3 verifies:

$$\frac{P_3P_1}{P_2P_3} = \frac{ns(P_2)}{ns(P_1)} \qquad (1)$$

This process is recursively repeated.

The above strategy may be generalized to 3D. Now the hemisphere is divided into four spherical triangles. Each spherical triangle is processed in the same manner: a ray is traced for each vertex from the center of the patch in order to determine the number of surfaces it intersects (ns(A) for vertex A). Then, on each edge, a new point is computed, using the above formula (1), which determines the direction of a new ray to shoot. These new points are obtained by the following relations:

$$\frac{AF}{FB} = \frac{ns(B)}{ns(A)} \qquad \frac{BD}{DC} = \frac{ns(C)}{ns(B)} \qquad \frac{CE}{EA} = \frac{ns(A)}{ns(C)}$$

These three new points, together with the initial vertices of the spherical triangle, define 4 new spherical triangles, which are processed in the same manner (Fig. 6.10).

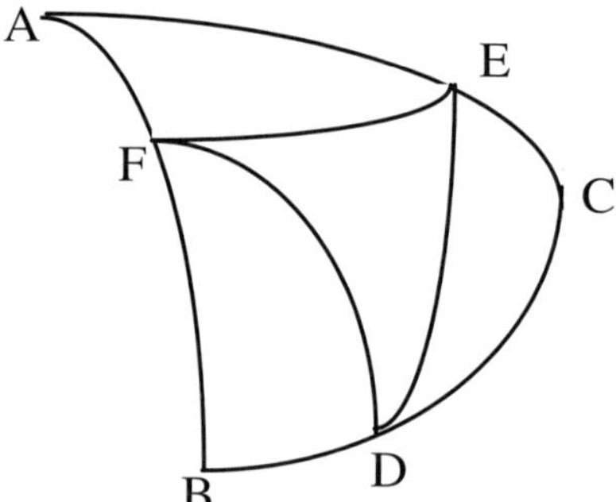

Fig. 6.10: Computation of 4 new spherical triangles

In practice, this naive hemisphere subdivision method is very expensive in memory occupation because it is difficult to store all the spherical triangles at each iteration. Indeed, for each patch, we need to store a great number of directions, depending on the subdivision level of spherical triangles. If n is the subdivision level, the number of spherical triangles to store for each patch will be 4^n for the whole hemisphere. Thus, for a number of patches of about 5000 and n=4, the number of spherical triangles to store for the whole scene will be equal to 5000 x 256 = 1 280 000, that is, about 2 000 000 vertices, taking into account that some vertices are double or triple. As, for each vertex, four bytes must be used, the algorithm needs 6 000 000 bytes of RAM for a weak subdivision level. For an additional subdivision step, the memory occupation will be of 24 000 000 bytes.

So, the implemented algorithm is slightly different and works as follows:

1. For each patch, rays are traced from a small number of points of the patch, chosen by a fictive regular subdivision of the patch, using the hemisphere subdivision method described in section 4. A hemisphere is computed for each point (Fig. 6.11).

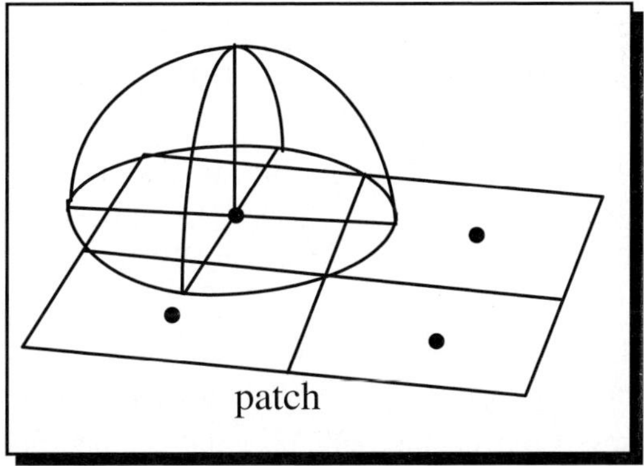

Fig. 6.11: Fictive patch subdivision and hemisphere computed for each sub-patch center

After each spherical triangle refinement, the same number of additional rays is traced randomly to each spherical triangle.
This hemisphere subdivision is pursued during a number n of steps given by the user. The obtained spherical triangles are stored at each step, in order to be subdivided in the following steps.

2. During another number m of steps, also given by the user, the same subdivision process is applied but now the spherical triangles are not stored. They are approximately estimated at each refinement step. The approximate estimation

process works as follows, where ens (A) means "estimated number of surfaces encountered in direction A" :

(i)   Initialise estimated numbers of surfaces for directions A, B and C :
       ens(A) = ns(A), ens(B) = ns(B), ens(C) = ns(C).

(ii)   At any subsequent refinement step, compute, from the current spherical tri-
       angle ABC, estimated values for new directions D, E and F. The following
       formulas give the way in which the estimated position of the new point F,
       and the estimated number of visible surfaces from this point are computed:

$$\frac{AF}{FB} = \frac{ens(B)}{ens(A)} \quad ens(F) = \frac{ens(A)\ BF + ens(B)\ AF}{AB}$$

3. Traditional Monte-Carlo progressive refinement is applied until a useful image
    is obtained.

    Rays shot towards a spherical triangle transport an amount of energy propor-
    tional to the area of this triangle.

6.3.3   Evaluation of the first hemisphere subdivision method

Comparison of the improved version of the hemisphere subdivision method with the Monte-Carlo progressive radiosity method shows that the former is much more efficient in terms of convergence. Fig. 6.12 and Fig. 6.13 show the obtained image for a scene with the two methods, after respectively one and five steps.

Table 1 compares convergence of the last version of the hemisphere subdivision method and the Monte-Carlo progressive radiosity method for the scene of Fig. 6.12 and Fig. 6.13. We can see that the hemisphere subdivision method converges faster than the Monte-Carlo progressive radiosity method.

 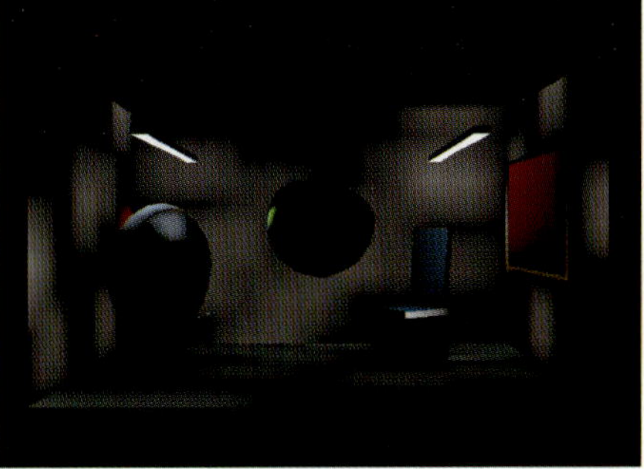

Fig. 6.12: Evaluation of the hemisphere subdivision method after one step. On the left, Monte Carlo progressive radiosity; On the right, hemisphere subdivision method (Images Vincent Jolivet)

METHOD NB OF STEPS	Monte-Carlo progressive radiosity	Hemisphere subdivision
1	0.43	0.40
2	0.39	0.36
5	0.15	0.09

Table 1: Comparing convergence of the two methods

Processing times are not given in table 1 because they are identical for the two methods. On the other hand, the hemisphere subdivision method is memory consuming, unlike the Monte-Carlo progressive radiosity method. However the cost in memory of the last version of the hemisphere subdivision method is not very inconvenient because it can be controlled by the user and remains stable after a few number of steps.

 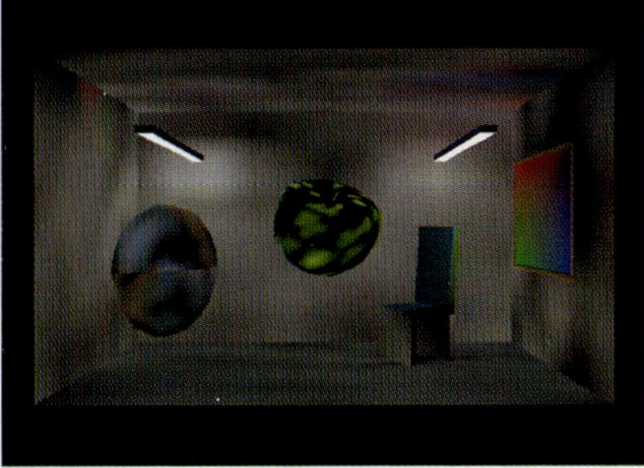

Fig. 6.13: Evaluation of the hemisphere subdivision method after five steps. On the left, Monte Carlo progressive radiosity; On the right, hemisphere subdivision method (Images Vincent Jolivet)

6.3.4   A new approach of hemisphere subdivision

The hemisphere subdivision technique presented above is very interesting because it allows intelligent, viewpoint complexity-based, selection of rays shot from a patch during the radiosity computation process. All the tests we have made with the Feda's and Purgathofer's algorithm confirm the improvements due to this technique.

However, we have seen that the brute hemisphere subdivision technique is a memory consuming technique and so, this technique cannot be applied to scenes divided into an important number of patches. For this reason, in the version of the hemisphere subdivision technique really implemented, the subdivision of spherical triangles becomes fictive after a small number of initial subdivisions and the computation of the scene density at vertices of fictive spherical triangles is an approximate one.

The purpose of the new approach presented in this section is to make the hemisphere subdivision technique as accurate as possible, with a memory cost allowing to process complex scenes.

The main principles which guide this approach are the following:

- Separation of the role of rays. The rays shot to determine regions do not distribute energy.

- Separation of the subdivision process from the radiosity computing process.

- Increase of the accuracy of the technique by avoiding fictive triangle subdivisions and approximate computations of scene densities.

- Decrease of the memory cost by reducing the maximum number of subdivisions of a spherical triangle.

After having presented the main purposes of the new hemisphere subdivision method, let us describe the method more precisely.

6.3.4.1 Preprocessing

A hemisphere, divided into 4 initial spherical triangles, is associated with each patch of the scene (Fig. 6.14).

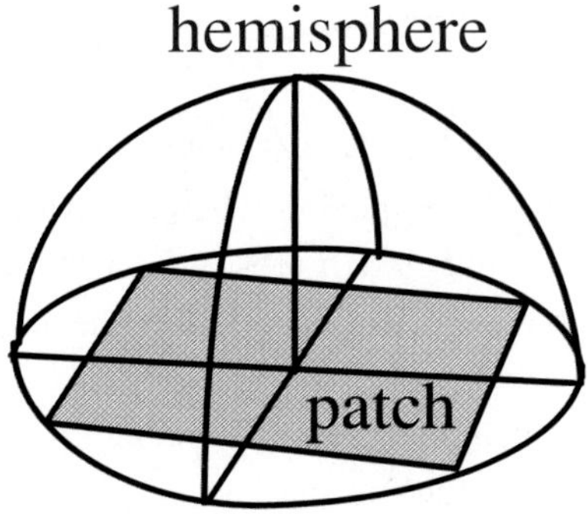

Fig. 6.14: Initial subdivision of a hemisphere

Each hemisphere is processed in the following manner:

6   From each centre of patch, three rays are shot to each spherical triangle, one
    ray per vertex, in order to determine the scene's density in the direction de-
    fined by the centre of the patch and the vertex (Fig. 6.15).

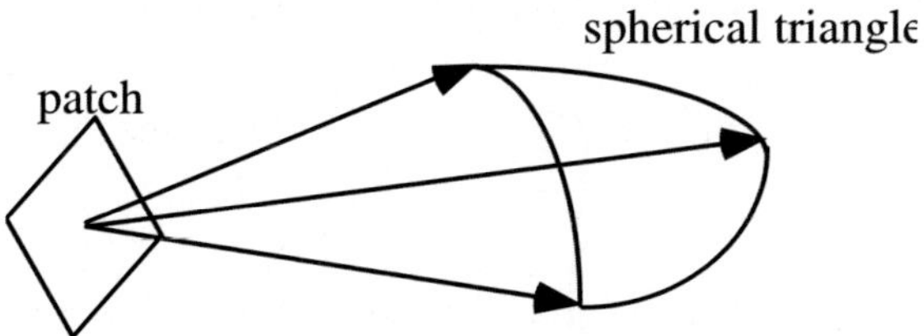

Fig. 6.15: One ray is sent to each vertex of the triangle

2. If the average value of densities for a spherical triangle is higher than a threshold value, the spherical triangle is divided into four new equal spherical triangles. In Fig. 6.16 a spherical triangle is divided into four new spherical triangles while in Fig. 6.17 adaptive subdivision of a spherical triangle is shown at
   the end of the preprocessing phase.

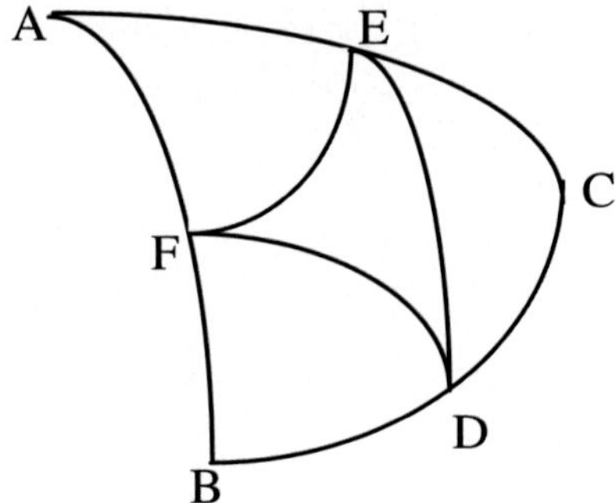

Fig. 6.16: The initial spherical triangle ABC is divided into 4 equal new spherical
triangles

Actually, the maximum subdivision depth used for the preprocessing phase is
equal to 2 but it can be changed interactively. So, each initial spherical triangle of
each hemisphere can be subdivided into a maximum of 16 spherical triangles.

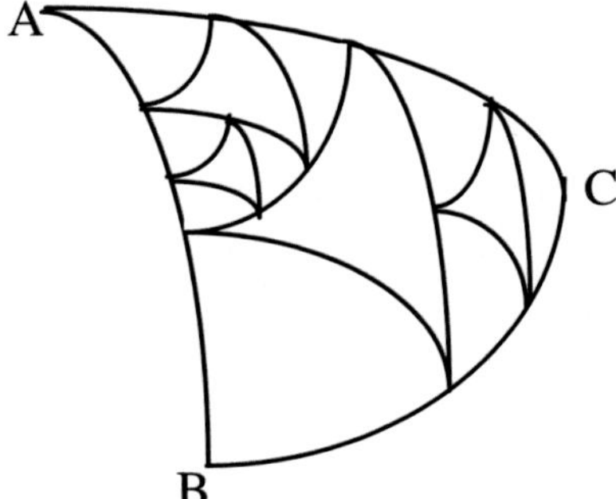

Fig. 6.17: Selective subdivision of a spherical triangle

6.3.4.2 Radiosity computation

This phase works during any number of steps and can be stopped if a useful image
is obtained.

1. From the center of each patch of the scene, a number of additional rays are
    shot towards the remaining of the scene. A number of rays proportional to the
    average value of its density (viewpoint complexity) are shot in each spherical
    triangle of the hemisphere, towards directions chosen according to a cosine
    distribution. The amount of energy distributed by each ray shot to a given

spherical triangle is the same, proportional to the area of the spherical triangle and inversely proportional to the number of rays shot in the triangle.

2. The radiosity of each patch computed up until the previous step is combined with the radiosity computed in the current step, in order to compute the radiosity of the patch up until the current step.

The general algorithm of the new hemisphere subdivision technique for computing radiosity is briefly presented here:

Procedure NewHemisphereRadiosity (Scene)
      Preprocessing (Scene)
      Compute Radiosity (Scene)
end

The two procedures Preprocessing and Compute Radiosity could be written as follows:

Procedure Preprocessing (Scene)
      for each patch **of** Scene **do**
        for each spherical triangle **of** Hemisphere (patch) do
          if density (spherical triangle) >ThresholdValue
                and SubdivisionLevel < MaxLevel **then**
            Divide (Spherical Triangle)
        end
      end
end

In the procedure Compute Radiosity, the meaning of the used variables is the following:

Ntr, Nhemi: Number of rays to shoot in the spherical triangle and in the hemisphere.

Atr, Ahemi: Area of the spherical triangle and ot the hemisphere.

Dtr, Daverage: Density of the spherical triangle and average density of the hemisphere.

Ftr, Fpatch: Energy to shoot in the spherical triangle and total energy to send from the patch.

Fray: Energy sent in the spherical triangle by a ray.

Procedure Compute Radiosity (Scene)

    while not Useful (Image) **do**

        for each patch **of** Scene **do**

            for each spherical triangle **of** Hemisphere (patch) do

$$N_{tr} := N_{hemi} \; \frac{A_{tr}}{A_{hemi}} \; \frac{D_{tr}}{D_{average}}$$

$$\Phi_{tr} := \Phi_{patch} \; \frac{A_{tr}}{A_{hemi}}$$

$$\Phi_{ray} := \frac{\Phi_{tr}}{N_{tr}}$$

              Choose N_{tr} directions with cosine distribution

              for each direction **do**

                  Shoot a Ray (Direction, Φ_{ray})

              end

            end

        end

    end

end

6.3.5 Discussion and results

The new hemisphere subdivision method described in subsection 6.3.4 improves
the first hemisphere subdivision method presented in [17].

The main advantage of this method is to decrease the error committed in regions
comporting several details, while the error in regions with few details is not increased significantly. Indeed, the mean square error in regions comporting many
details is less than with a classical Monte-Carlo method because more rays are
sent towards these regions. In return, although the mean square error in regions
with few details is greater than with a classical Monte-Carlo method, this difference is hardly detected by the human eye. So, a useful image is generally obtained
very sooner than with the Feda's and Purgathofer's method.

The memory cost of the method is not very important, even for complex scenes. It
increases linearly with the complexity (in number of patches) of the scene.

Thus, if n is the maximum subdivision depth of a hemisphere and p the number of
patches of the scene, the maximum number t of spherical triangles we have to
process for a hemisphere is given by:

$$t = \frac{4 \, (4^{n+1} - 1)}{3}$$

For each spherical triangle, the only information we have to store is the average value of its density and this information can be stored in a byte. So, the total memory requirement is:

$$m = p\,t \;\;\text{bytes}$$

or

$$m = \frac{4\,p\,(4^{n+1}-1)}{3} \;\;\text{bytes}$$

Thus, for a complex scene composed of 100 000 patches, a storage capability of 6 400 000 bytes is required.

In Fig. 6.18 one can see an image of a simple scene processed with the classical progressive refinement Monte Carlo radiosity method after 3 steps, while in Fig. 6.19 we have an image of the same scene processed with the new hemisphere subdivision technique after 3 steps. The scene is composed of 1002 patches.

Fig. 6.18: image of a scene after 3 steps with progressive refinement Monte Carlo radiosity (Image Vincent Jolivet)

Fig. 6.19: image of a scene after 3 steps with our new hemisphere subdivision method (Image Vincent Jolivet)

The maximum number of spherical triangles of a hemisphere is 84 for a maximum subdivision level equal to 2.

In Fig. 6.20 we have an image of a more complex scene processed with the classical progressive refinement Monte Carlo radiosity method after 17 steps, while in Fig. 6.21 we have an image of the same scene processed with the new hemisphere subdivision technique after 17 steps.

This scene is composed of 3576 patches and the maximum number of spherical triangles used by the new hemisphere subdivision technique is 84.

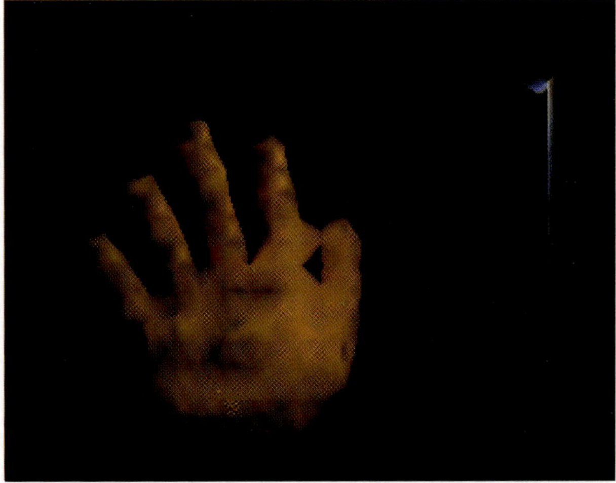

Fig. 6.20: image of a scene after 17 steps with progressive refinement Monte Carlo radiosity (Image Vincent Jolivet)

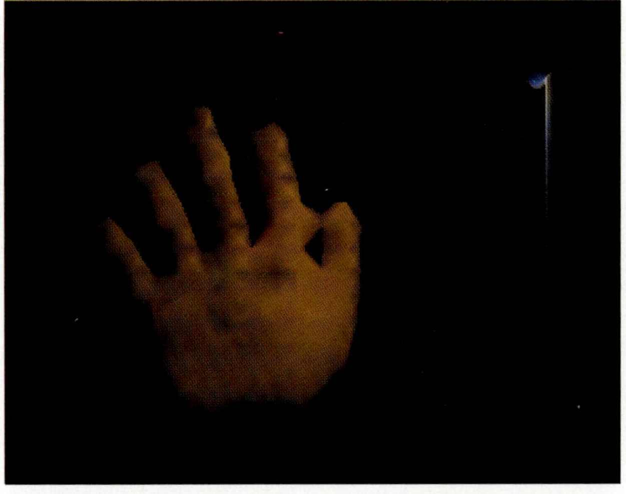

Fig. 6.21: image of a scene after 17 steps with the new hemisphere subdivision method (Image Vincent Jolivet)

6.3.6   Conclusion

The Hemisphere Subdivision Method is a very interesting importance driven method for computing radiosity. Its main drawback was its cost in memory because of the need to store a big quantity of information about spherical triangles. The new technique of the Hemisphere Subdivision Method presented here improves the initial method by significantly reducing the storage requirements.

The main purpose of the technique presented in this section is to improve the progressive refinement Monte-Carlo radiosity by using rough estimation of viewpoint complexity and hence optimizing the radiosity computation in the complex parts of the scene, in order to obtain a useful image of the scene very soon. The results of implementation of this technique show that it permits to obtain a useful image faster than the classical Monte-Carlo progressive refinement technique. However, this technique could possibly fail in some situations where a wall is in front of a large number of surfaces. But, even in such a case, the hemisphere subdivision technique is, at least, as fast as Monte-Carlo progressive radiosity technique.

6.4 Pyramidal subdivision

6.4.1 Pyramidal hemisphere subdivision

Results obtained with the new hemisphere subdivision method are rather interesting but it is clear that this method has two main drawbacks:

- It is necessary to clarify the function of rays shot in the scene. Some rays serve only to diffuse the scene energy while others have a double function: determine good directions and diffuse energy.

- All the used rays transport the same energy. With this choice the method could have convergence problems. So, rays shot towards a spherical triangle should transport a part of energy proportional to the area of this triangle.

The main idea of this technique is to avoid the use of rays in the good-direction-determining process. In this case, good directions will be computed globally by using a triangular pyramid. This pyramid is defined by the center of a patch and the three vertices of the current spherical triangle (see Fig. 6.22). Two variants of the method can be considered : adaptive pyramidal subdivision and regular pyramidal subdivision.

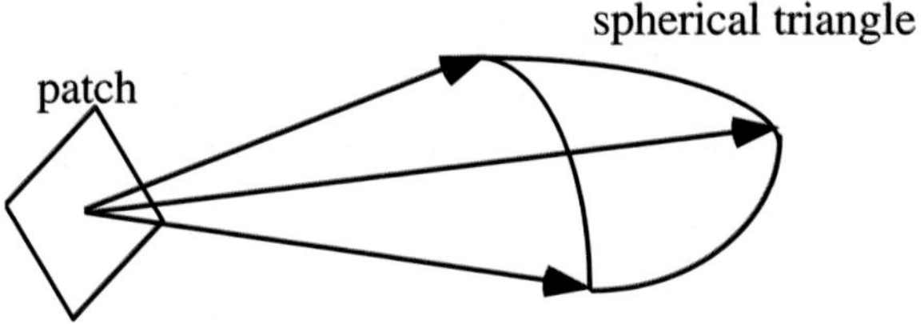

Fig. 6.22: Triangular pyramid associated with a spherical triangle

In both variants triangular surface patches are used, instead of rectangular ones, and, unlike in the hemisphere subdivision method, the hemisphere surrounding a patch is divided into three spherical triangles. In figure 6.23, the three spherical triangles are defined by the top T and arcs AB, BEC and ECD respectively.

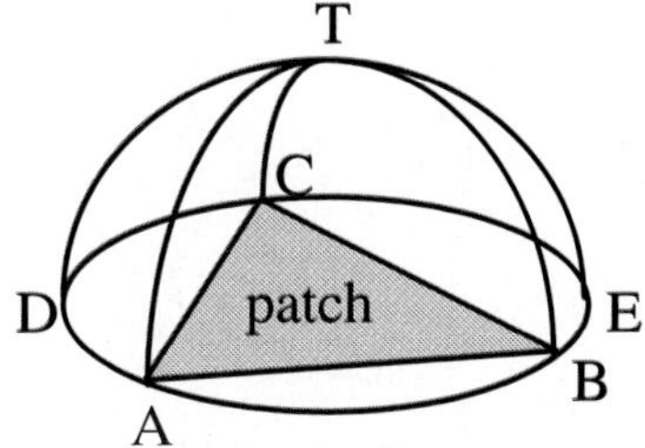

Fig. 6.23: Hemisphere subdivision into 3 spherical triangles

Like in the hemisphere subdivision technique, the good direction criterion, that is the viewpoint quality rough estimation, is the density of the scene in a direction, that is, the number of objects encountered in this direction. The choice of this criterion is justified by the fact that regions with a great number of objects are probably more complex than regions which contain a small number of objects. Of course, this is not always true but this is true very often and the use of the above criterion in the pyramidal hemisphere subdivision technique improves Feda's algorithm for all the tested scenes.

6.4.2 Adaptive Pyramidal Subdivision

In the hemisphere subdivision method, the hemisphere surrounding a surface patch is divided into four spherical triangles. In this new method, the hemisphere surrounding a surface patch is divided into three spherical triangles, because of the use of triangular patches. Each spherical triangle becomes the basis of a pyramid whose top is the center of the patch. Each pyramid is now processed in the same manner. Two phases are used in the processing of a pyramid : the pyramid subdivision phase and the ray tracing phase.

First phase: Pyramid subdivision.
If the current pyramid contains a number of objects greater than a threshold value, the corresponding spherical triangle is subdivided into four new triangles defined by the vertices of the current spherical triangle and the middles of its edges (see Fig. 6.24). A triangular pyramid is associated with each new triangle.

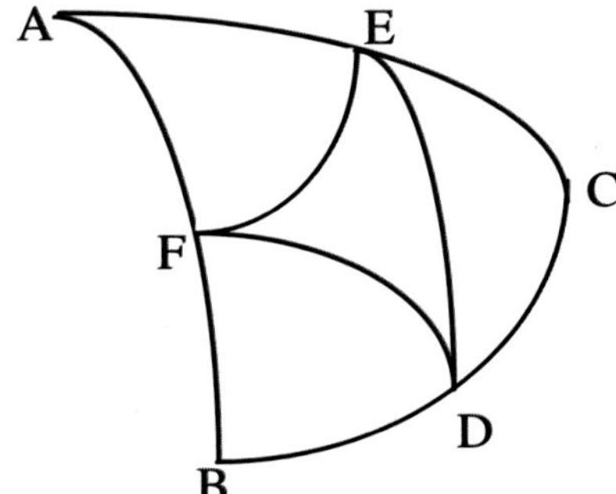

Fig. 6.24: The initial spherical triangle ABC is divided into 4 equal new spherical triangles

The same subdivision process is applied to each one of the obtained spherical triangles and its corresponding triangular pyramid.

Thus, each initial spherical triangle is selectively divided by a heuristic search process. Pyramids containing many objects are more refined than pyramids containing few objects (see Fig. 6.25).

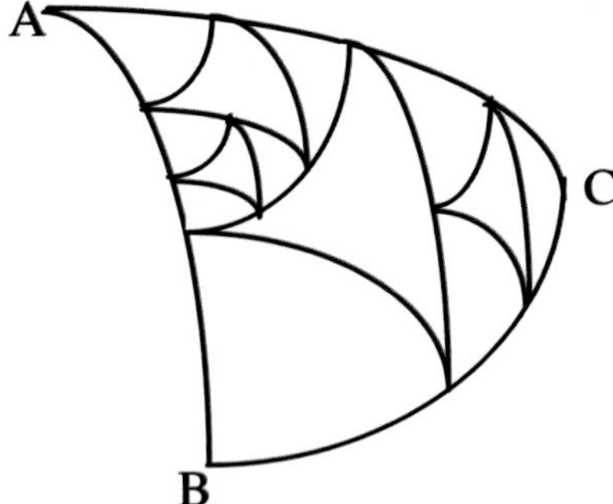

Fig. 6.25: Selective subdivision of a spherical triangle

Second phase: Ray tracing

The ray tracing phase includes an initial ray tracing process and a multi-step ray tracing process.

During the initial ray tracing process, the same small number of rays are randomly shot into each triangular pyramid from the center of the patch. The amount of energy transported by each ray is proportional to the spherical triangle area.

The multi-step ray tracing process is accomplished in several steps. At each step, the following actions are performed:

- A number of additional rays are randomly shot into each triangular pyramid in order to refine the obtained image. The rays shot are supposed to distribute the whole energy of the emitter patch. The amount of energy transported by each ray is proportional to the corresponding spherical triangle area.
- If the total number of rays shot up until the current step is k times greater than the number of rays shot up until the previous step, the energy received by each patch up until the previous step and the energy distributed by the current patch are divided by k and the remaining energy of the current patch is diffused by the additional rays shot in the current step.
- An image of the scene is displayed at the end of the step.

The ray tracing phase is achieved when the energy of the scene has been entirely distributed or when a useful image has been obtained.

6.4.3 Regular Pyramidal Subdivision

Like in adaptive pyramidal subdivision technique, the hemisphere surrounding a surface patch is divided into three spherical triangles. Each spherical triangle becomes the basis of a pyramid whose top is the center of the patch. Each pyramid is now processed in the same manner, in two phases.

First phase: Pyramid subdivision.

Each pyramid is regularly subdivided until a subdivision level which can be fixed or given by the user. Pyramid subdivision is performed by subdividing the corresponding spherical triangle into four new triangles defined by the vertices of the current spherical triangle and the middles of its edges. A triangular pyramid is associated with each new triangle.

If the subdivision level is not reached, the same process is applied to each one of the obtained spherical triangles and its corresponding triangular pyramid.

Thus, each initial spherical triangle is regularly divided by a normal subdivision process. Each triangular pyramid contains its own number of objects (see Fig. 6.26).

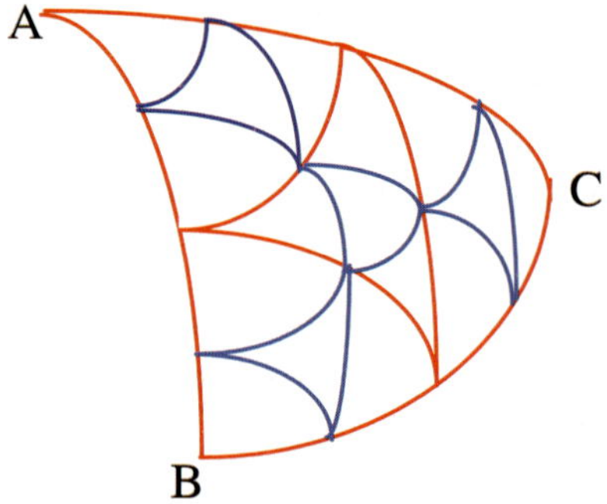

Fig. 6.26: Regular subdivision of a spherical triangle

Second phase: ray tracing.
Like with the adaptive subdivision technique, the ray tracing phase includes an initial ray tracing process and a multi-step ray tracing process.
During the initial ray tracing process, a small number of rays, proportional to the number of objects contained into the spherical triangle area, are randomly shot into each triangular pyramid from the center of the patch. The amount of energy transported by each ray is now inversely proportional to the number of rays shot into the spherical triangle area.
The multi-step ray tracing process is accomplished in several steps. At each step, the following actions are performed:
- Additional rays are randomly shot into each triangular pyramid in order to refine the obtained image. The rays shot are supposed to distribute the whole energy of the emitter patch. The amount of energy transported by each ray is inversely proportional to the number of the additional rays shot.
- If the total number of rays shot up until the current step is k times greater than the number of rays shot up until the previous step, the energy received by each patch up until the previous step and the energy distributed by the current patch

are divided by k and the remaining energy of the current patch is diffused by the additional rays shot in the current step.

- An image of the scene is displayed at the end of the step.

The ray tracing phase is achieved when the energy of the scene has been entirely distributed or when a useful image has been obtained.

6.4.4 Number of Objects Contained Into a Pyramid

The computation of the number of objects contained into a triangular pyramid is necessarily approximate, for two reasons. The first reason is that it is not always easy to compute the exact number of objects contained into a pyramid. The second reason is that this computation must be as fast as possible in order not to penalize the radiosity algorithm.

An object is considered to be entirely inside a pyramid if all vertices of the object are contained in the pyramid. An object is considered to be partially inside a pyramid if a vertex of the object is contained into the pyramid. Three possibilities could be considered:

1. Only objects belonging entirely to the pyramid are considered to be inside the pyramid and the object counter is incremented by one for each one of these objects (see Fig. 6.27). The number of objects n considered to be inside the pyramid is given by:

   n = e

   where e is the number of objects entirely contained into the pyramid.

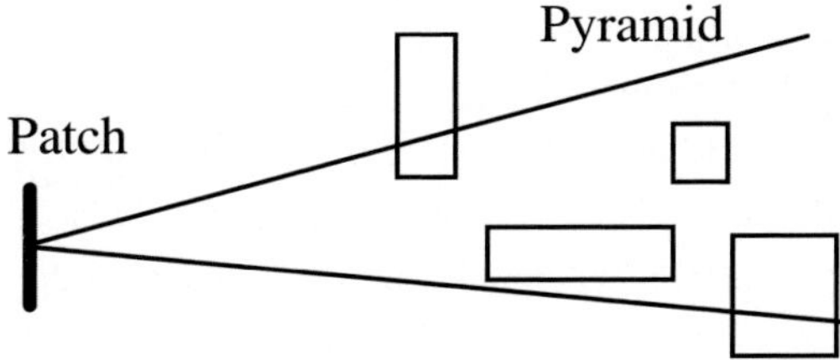

Fig. 6.27: Two objects are considered to be inside the pyramid

2. All objects, entirely or partially contained into the pyramid, are considered to be inside the pyramid. The object counter is incremented by one for each object entirely contained into the pyramid and by a half for each object partially inside the pyramid (see Fig. 6.28). The number of objects n considered to be inside the pyramid is given by the formula :

$$n = e + \frac{1}{2} p$$

where e is the number of objects entirely contained into the pyramid and p the number of objects partially contained into it.

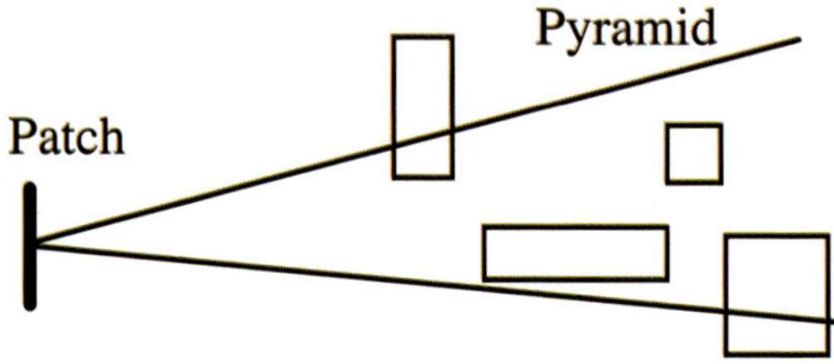

Fig. 6.28: Three objects are considered to be inside the pyramid

3. All objects, entirely or partially contained into the pyramid, are considered to be inside the pyramid. The object counter is incremented by one for each object entirely or partially inside the pyramid (see Fig. 6.29). The number of objects n considered to be inside the pyramid is given by the formula :

$$n = e + p$$

where e is the number of objects entirely contained into the pyramid and p the number of objects partially contained into it.

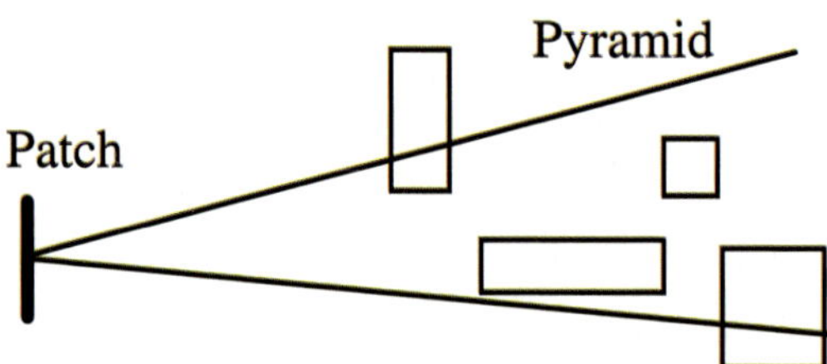

Fig. 6.29: Four objects are considered to be inside the pyramid

The third solution has been chosen because it is much faster than the other solutions. Indeed, with this solution it is sufficient to find a vertex of an object in the pyramid to decide that the object is inside the pyramid.

6.4.5 Discussion and Results

The main advantage of the adaptive pyramidal subdivision is that subdivision is rather intelligent because it takes into account the number of objects contained in a region. Its main drawback is that this technique must maintain different levels of subdivision and, so, its need of memory space is very important.

The main advantage of the regular pyramidal subdivision is the simplicity of the management of spherical triangles subdivision because all spherical triangles of all hemispheres are subdivided at the same level. Thus, the quantity of information to store is very small because the same hemisphere can be used for each patch of the scene, after a simple geometric transformation (translation or/and rotation). The main drawback of this technique is that subdivision is not driven by the density of the scene and useless subdivisions can be applied to regions containing a small number of objects.

Estimation of a global complexity for a region from a patch, based on the number of objects contained in the region, is more accurate than estimation of local complexities and approximation of a global complexity by computing the average value of local complexities. However, the main drawback of these techniques for estimating the visual complexity of a region is that the complexity estimated by both techniques is not a visual one. Indeed, computing the number of objects lying in a direction or in a region only permits to estimate the real complexity of a region and this real complexity is not the viewpoint complexity of the region. In many cases, the real complexity of a region of the scene can be used to estimate its viewpoint complexity but this estimation can fail, especially in cases where a big object hides several objects lying behind it. Thus, in the case where a wall hides several parts of the scene, both techniques used to estimate the viewpoint complexity of a region, fail.

In order to compute a more accurate viewpoint complexity, an improvement was introduced in the technique based on the number of objects (patches) contained in a region. This improvement takes into account the number of patches contained in the pyramid and receiving energy from the current patch.

Let us consider Fig. 6.30, where 6 objects are contained in the pyramid associated with the current patch. Among all these objects, only one, the object no 1, receives energy from the rays shot from the current patch. This fact permits to conclude that it is useless to shoot a great number of rays in the pyramid since only a small percentage of the objects contained in the pyramid is concerned by these rays.

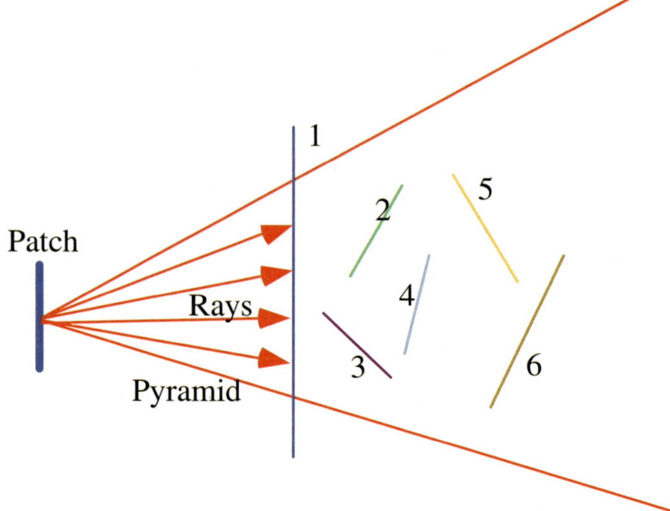

Fig. 6.30: The number of patches receiving energy from the emitting patch permits to dynamically improve the rays distribution.

So, at each step of energy distribution, the ratio: number of objects in the pyramid which received energy during the last step / number of rays sent in the pyramid is computed and serves to improve the rays distribution in the next step. The number of rays sent in the pyramid is dynamically updated at each energy distribution step, this number increasing or decreasing with the value of the above ratio.

Fig. 6.31: Detail of a reference scene obtained with classical Monte Carlo radiosity calculation (left image) and global density-based radiosity calculation (right image) at the end of the first step. (Images Vincent Jolivet)

This improvement greatly increases the efficiency of the criterion estimating the visual complexity of the scene. Fig. 6.31 shows results of radiosity calculation using this technique. However some drawbacks of the estimation technique remain, as it will be explained in the next subsection.

6.4.6   How to improve viewpoint complexity estimation

The above improvement of pyramidal subdivision technique greatly increases the efficiency of the criterion estimating the visual complexity of the scene. Figure 9 shows results of radiosity calculation using this technique. However some drawbacks of the estimation technique remain:

1. Even if the viewpoint complexity is progressively approximated, the hemisphere subdivision is based only on real complexity criteria because the above improvement is not applicable during the subdivision step.

2. The objects considered by the used criteria to estimate the visual complexity of a scene are patches. Unfortunately, the estimated number of visible patches cannot give a precise idea of the visual complexity of a region of the scene. A region containing a single flat surface divided into n visible patches, is generally much less complex than a region containing several small surfaces, even if the number of visible patches is less than n. In Fig. 6.32, two regions of a scene are considered from the current patch, that is, the patch which has to distribute its energy in the scene. In both regions, the number of visible patches is 5 but it is easy to see that the viewpoint complexity of the second region is greater than the viewpoint complexity of the first one. A small number of rays is sufficient to distribute the corresponding part of energy of the current patch in the first region while a more important number of rays is necessary to accurately distribute energy in the second region.

So, it becomes clear that the more accurate criterion of visible complexity of a region is the number of surfaces seen in the region. The question is: how to approximate the number of visible surfaces in a scene without increasing the complexity of radiosity computation?

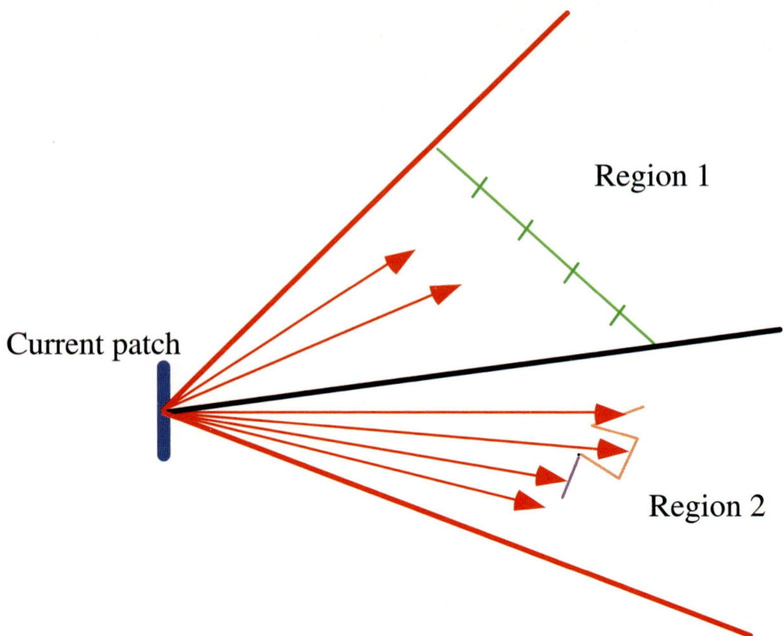

Fig. 6.32: the number of visible patches is not always a good viewpoint complexity criterion

In the following section, a new technique to estimate the viewpoint complexity
of a region of a scene is presented. This technique avoids the two main drawbacks
of the criteria currently used by the hemisphere subdivision methods to compute
radiosity.

6.5 More accurate Monte Carlo radiosity

We have seen that the criteria used for sampling the scene as seen from a patch are
not always very precise and can fail in some cases. Even with the improvement
described in section 3, the solution is not fully satisfactory because the improvement occurs after the scene has been sampled and because the criterion used in
this improvement - estimation of the number of visible patches - is not precise
enough. In order to avoid these drawbacks, a new scene and rays sampling technique will be proposed in this section, based on adaptive refinement of a region.

Adaptive refinement technique can be used to improve the radiosity computation process for Monte Carlo based methods. The purpose of this technique is to
introduce a new viewpoint complexity criterion estimation for the hemisphere

subdivision method, the number of visible surfaces. The expected improvement is an enhanced processing of complex parts of the scene. Two variants of this technique can be imagined. In the first one, subdivision of a region depends on the state of neighboring regions while in the second variant regions are processed independently of each other.

6.5.1 Subdivision depending on the neighboring regions

This technique uses a region subdivision criterion based on the state of neighbouring regions. The technique will be first explained in 2 dimensions, in order to be easier to understand. We will see that transposition to 3 dimensions is not always easy.

A hemisphere is associated to each patch of the scene. The technique works in two phases, preprocessing and radiosity computation.

6.5.1.1 Preprocessing

1.   The hemisphere is regularly subdivided into a fixed number of zones (Fig. 6.33).

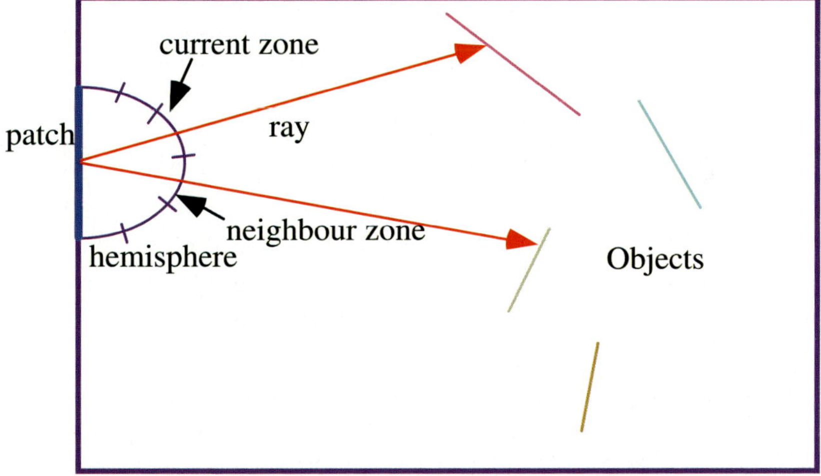

Fig. 6.33 : Selective scene refinement for a hemisphere

2.   At each step, a ray is sent to the center of each zone from the center of each patch (or from a randomly chosen point of the patch).
      If the visible surface (not the visible patch) for the current zone is different from the visible surface for a neighbor zone, the current zone is subdivided and the same process is applied to each resulting zone.
3.   The whole process stops when a given subdivision level is reached.

6.5.1.2 Radiosity computation

At each step, the same number of additional rays is shot against each zone. The power transported by each ray is proportional to the area of each zone.

The radiosity computation process can be stopped at any step.

This variant is difficult to transpose in 3 dimensions, as the zones are now spherical triangles and it is difficult to find the neighbors of a spherical triangle (Fig. 6.34). Implementation of this variant supposes a heavy organization of spherical triangles in order to find, as easily as possible, the neighbors of a given spherical triangle.

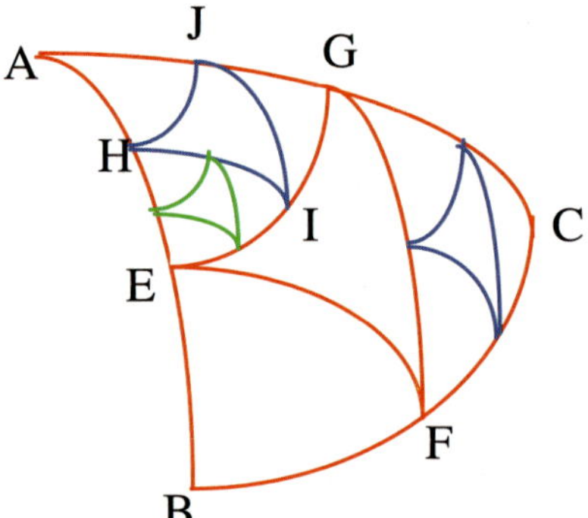

Fig. 6.34 : Which are the neighbors of the triangle IGJ ?

In order to give a solution to the neighbor finding problem, another variant of the selective refinement technique is proposed.

6.5.2 Independent processing of regions

A hemisphere is associated to each patch of the scene. The variant works also in two phases, preprocessing and radiosity computation. Only the preprocessing phase is different from the one of the first variant.

6.5.2.1   *Preprocessing*

1.   The hemisphere is regularly subdivided into a fixed number of zones (Fig. 6.35).
2.   At each step, a ray is sent to each extremity of each zone from the center of each patch (or from a randomly chosen point of the patch).

     If the visible surface (not the visible patch) for the current zone is not the same for all the extremities of the zone, the zone is subdivided and the same process is applied to each resulting zone.

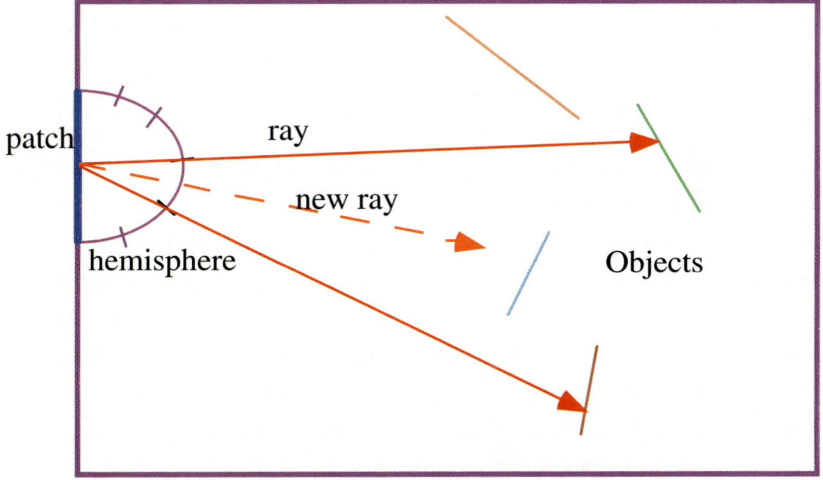

Fig. 6.35: a new ray is to be traced

3.   The whole process stops when a given subdivision level is reached.

   This variant is easy to transpose in 3 dimensions. Now, a zone is a spherical triangle and a ray is shot against each vertex of the triangle. The spherical triangle is subdivided into 4 new spherical triangles (see Fig. 6.36).

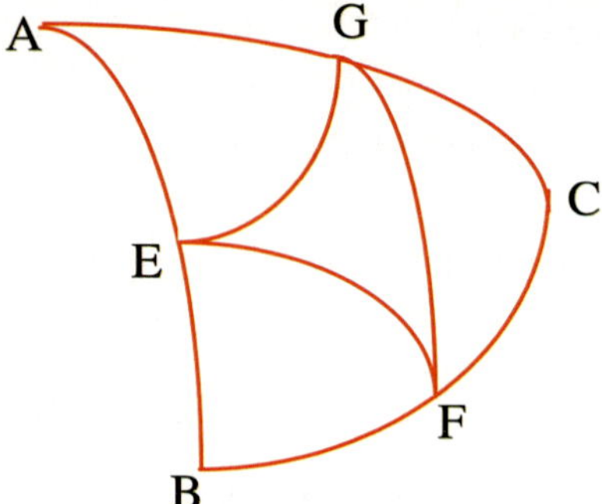

Fig. 6.36: subdivision of a spherical triangle

6.5.2.2 Radiosity computation

At the end of the subdivision process, the same number of rays is shot to each spherical triangle. The energy transported by each ray is proportional to the area of the triangle.

6.5.3 Discussion

The main advantage of the techniques proposed in this section is that subdivision is performed whenever changes, concerning the visible surface, are observed between two parts of a region. The approximate number of visible surfaces in a region, that is the number of changes of visible surface, is a good criterion of the viewpoint complexity of a scene from a patch, because a region with many changes of visible surface must be processed more accurately than a region with few changes. The only drawback of these techniques is that the visible surface must be computed for each ray shot during the hemisphere subdivision (sampling) phase.

6.6   Conclusion

The techniques presented in this chapter try to exploit the knowledge of a scene in order to perform efficient computation of radiosity. Unlike other importance-driven techniques, the techniques developed here use heuristic search and viewpoint complexity estimation in order to delimit important regions. A detailed computing of radiosity is then performed for these regions while the radiosity of less important regions is computed less accurately.

The first three families of methods presented in this chapter have been implemented and the obtained results are actually very satisfactory. Both, the last version of the hemisphere subdivision technique and the hemisphere pyramidal subdivision give good results, which improve the efficiency of existing algorithms by permitting a faster computation of a useful image and final computation of a more accurate image.

The results obtained with the techniques described in this chapter are always interesting and improve Monte Carlo based radiosity with all tested scenes.

A new, more accurate, sampling technique has been proposed in section 6.5, which should improve results of Monte Carlo-based radiosity computation. However, this technique was not yet implemented. A possible drawback of this technique is it's cost in computing time, because visibility has to be computed for each ray sent.

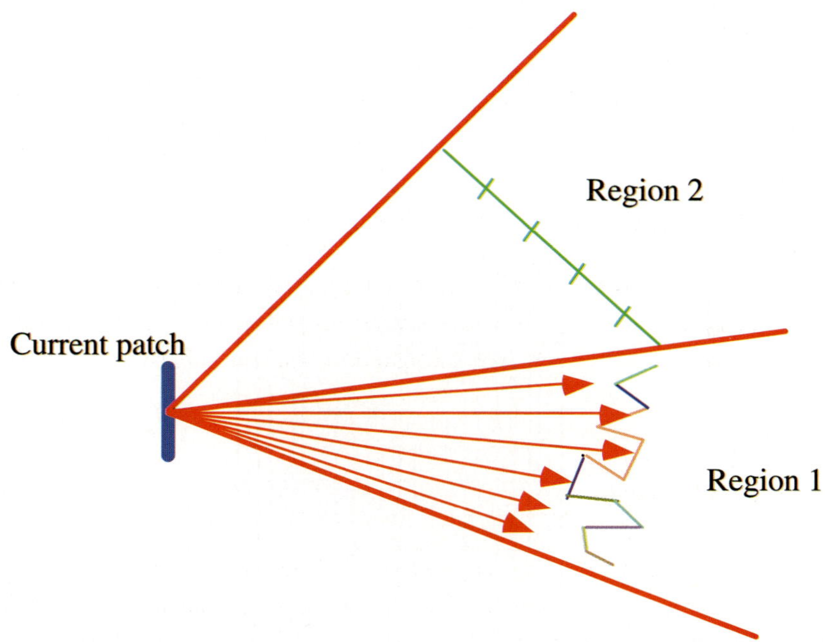

Fig. 6.37: If a region is much more complex than the other one, all available rays could be shot in the first region

From a theoretical point of view, it is possible that the sampling techniques described in the previous sections of this chapter fail, because of an insufficient number of rays to be sent. For example, if the number of rays to shoot in a scene composed of two regions is 1000, if the visual complexity of the first region is 10000 times greater than the visual complexity of the second one, all 1000 rays will be sent in the first region and no energy will be diffused in the second region (Fig. 6.37).

In reality, this case is difficult to occur because the subdivision process is stopped at a stage where the scene is subdivided in a relatively small number of regions and the ratio between the visual complexities of two regions is not very important. In any case, it is possible to decide that only a percentage (say 80%) of rays are shot according to the visual complexity criterion, the remaining rays being shot according the traditional Monte Carlo radiosity. Another solution is to take also into account the number of patches contained in a region to compute the number of rays to shoot in this region.

In all methods, roughly estimated region complexity is used in order to have a fast estimation and not to penalize the radiosity algorithm. However, we think that it is possible to have a fast, more accurate estimation of the visual complexity of a region from a patch, based on the method used in chapter 1 for fast accurate viewpoint complexity estimation. The method includes the following steps:

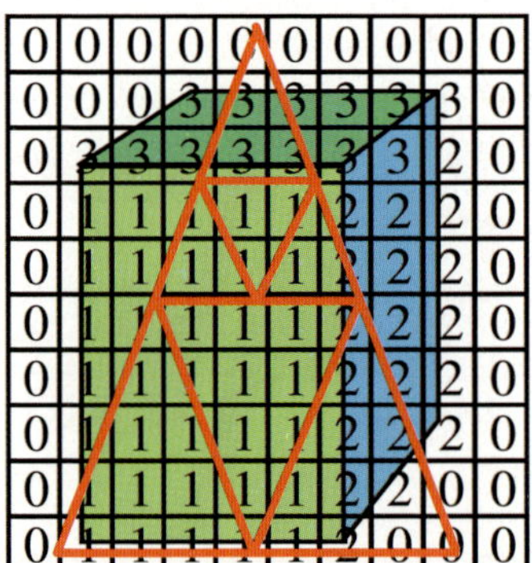

Fig. 6.38: accurate estimation of the visual complexity of a region

1. Give a different colour to each polygon of the scene.
2. Display the scene using the integrated z-buffer of OpenGL.
3. Estimate the visual complexity of the region, using the part of the image covered by the projection of the current spherical triangle (Fig. 6.38).
4. Subdivision, according on the estimated visual complexity.

References

1. Cohen, M., Greenberg, D.P., Immel, D.S., Brock, P.J.: An efficient radiosity approach for realistic image synthesis. IEEE Computer Graphics and Applications 6(3), 26–35 (1986)
2. Cohen, M., Chen, S.E., Wallace, J.R., Greenberg, D.P.: A progressive refinement approach to fast radiosity image generation. Computer Graphics (SIGGRAPH 1988 Proc.) 22(4), 75–84 (1988)
3. Colin, C.: Automatic computing of scene "good views". In: MICAD 1990, Paris (February 1990) (in French)
4. Feda, M., Purgathofer, W.: Progressive Ray Refinement for Monte Carlo Radiosity. In: Fourth Eurographics Workshop on Rendering, Conf. Proc., pp. 15–25 (June 1993)
5. Goral, G.M., Torrance, K.E., Greenberg, D.P., Battaile, B.: Modeling the Interaction of Light Between Diffuse surfaces. Computer Graphics (SIGGRAPH 1984 Proc.) 18(3), 212–222 (1984)
6. Hanrahan, P., Salzman, D., Aupperle, L.: A rapid hierarchical radiosity algorithm. Computer Graphics (SIGGRAPH 1991 Proc.) 25(4), 197–206 (1991)
7. Kamada, T., Kawai, S.: A Simple Method for Computing General Position in Displaying Three-Dimensional Objects. Computer Vision, Graphics and Image Processing 41 (1988)
8. Pattanaik, S.N., Mudur, S.P.: Computation of global illumination by Monte Carlo simulation of the particle model of light. In: Third Eurographics Workshop on Rendering, Conf. Proc., May 1992, pp. 71–83 (1992)
9. Plemenos, D.: A contribution to study and development of scene modeling, generation et display techniques - The "MultiFormes" project. Professorial dissertation, Nantes, France (November 1991) (in French)
10. Plemenos, D., Pueyo, X.: Heuristic sampling techniques for shooting rays in radiosity algorithms. In: 3IA 1996 Conference, Limoges, France, April 3-4 (1996)
11. Shirley, P.: Radiosity Via Ray Tracing. In: Arvo, J. (ed.) Graphics Gems II, pp. 306–310. Academic Press, San Diego (1991)
12. Wallace, J.R., Elmquist, K.A., Haines, E.A.: A ray Tracing Algorithm for Progressive Radiosity. Computer Graphics (SIGGRAPH 1989 Proc.) 23(3), 315–324 (1989)
13. Orti, R., Puech, C.: Efficient computation of 2D form factors applied to dynamic environments. In: Annual conference of the AFIG, Marseille, pp. 223–230 (November 1995) (in French)
14. Plemenos, D.: Selective refinement techniques for realistic rendering of three dimensional scenes. Revue Internationale de CFAO et d'Infographie 1(4) (1987) (in French)
15. Jansen, F.W., Van Wijk, J.J.: Previwing techniques in raster graphics. In: EUROGRAPHICS (September 1994)
16. Cohen, M.F., Greenberg, D.P.: The hemi-cube, a radiosity solution for complex environments. In: SIGGRAPH (1985)
17. Jolivet, V., Plemenos, D.: The Hemisphere Subdivision Method for Monte-Carlo Radiosity. In: GraphiCon 1997, Moscow, May 21-24 (1997)
18. Durand, F., Drettakis, G., Puech, C.: The 3D visibility complex: a new approach to the problems of accurate visibility. In: Eurographics Rendering Workshop, Porto (1996)
19. Neumann, A., Neumann, L., Bekaert, P., Willems, Y.D., Purgathofer, W.: Importance-driven Stochastic Ray Radiosity, the problems of accurate visibility. In: Eurographics Rendering Workshop, Saint-Etienne (1997)

20. Jolivet, V., Plemenos, D., Sbert, M.: A pyramidal hemisphere subdivision method for Monte Carlo radiosity. Short paper. In: International Conference EUROGRAPHICS 1999, Milan, Italy, September 7-11 (1999)
21. Jolivet, V., Plemenos, D., Sbert, M.: Pyramidal hemisphere subdivision radiosity. Definition and improvements. In: International Conference WSCG 2000, Plzen, Czech Republic, February 7-11 (2000)
22. Plemenos, D.: Visual Complexity-Based Heuristic Sampling Techniques For The Hemisphere Subdivision Monte Carlo Radiosity. Internet Journal Computer Graphics & Geometry 8(3) (2006)

7 Scene understanding and human intelligence

Abstract. In this chapter we present some ideas and first results for efficient scene understanding in particular cases, where traditional visual understanding techniques are not sufficient or not possible. So, a technique is presented first, permitting to understand scenes containing mirrors, transparencies or shadows and, for this reason, difficult to understand with methods presented in previous chapters, computing a good viewpoint or exploring a scene. The presented technique uses object-space based contour drawing, in order to add apparent contour drawing, for real only objects of the scene, on rendered image. The second technique presented in this chapter faces the case where it is necessary, for understanding a scene, to have a look inside and outside the scene, but visual exploration is not possible, because the image of the scene must be printed. In such cases, the presented technique allows to add holes to the scene, in order to see simultaneously the outside of the scene and selected objects inside it.

Keywords. Scene understanding, Contour extraction, Visibility, Selective refinement, Photo-realistic rendering, Non photo-realistic rendering, Reflections, Shadows.

7.1 Introduction

In previous chapters we have seen how artificial intelligence techniques may be used to improve various computer graphics areas. However, there are cases where human intelligence may be used to resolve some practical problems in computer graphics, especially in scene rendering. The purpose of this chapter is to present cases where particular scene understanding problems may be resolved, or attenuated, by alternative visualization techniques. This is a kind of data visualization, applied to computer graphics.

In chapters 3 and 4 of this volume we have presented a lot of techniques allowing a good visual understanding of simple or complex scenes. These techniques are based on automatic selection of good viewpoints, and in the case of complex scenes (or virtual worlds), on automatic selection of a good camera path, permitting to well explore the current world.

D. Plemenos and G. Miaoulis: Visual Complexity, SCI 200, pp. 221–246.

All these techniques were tested with various more or less complex scenes and give good results. However, there are some cases where these techniques are not entirely satisfactory. This problem may be due to some particular conditions or constraints.

So, a first case where scene understanding based on automatic good viewpoint selection, or on automatic exploration, is not satisfactory, is due to the nature of scenes to be understood. When a scene contains lights, mirrors or transparent objects, it is sometimes difficult to understand it from its rendered image, even if a good point of view was automatically computed and even if automatic scene exploration is proposed. The main problem with such scenes is that it is difficult, and sometimes impossible, for the human eye to distinguish among real objects and illusions due to shadows, reflections or refractions. So, we have to invent other techniques to facilitate scene understanding for this kind of scenes [39, 40].

Another case is the one, where usual scene understanding techniques cannot be applied, due to particular constraints. Suppose, for example, that we have to describe and to illustrate in newspaper the new model of an aircraft. We must show the outside, as well as the inside of the aircraft. Visual exploration in such a case is impossible. If we must illustrate it with only one picture, we have to invent an intelligent way to give sufficient visual knowledge about the aircraft with this picture [39].

The main part of this work is divided into 3 sections. In section 7.2, a brief presentation of the main scene understanding techniques will be made. A lot of these techniques are based on automatic computation of a good view or of a good path for a virtual camera exploring the scene. As the purpose is to enhance the visualization by extracting apparent contours, we are going to complete this section with the study the existing methods in this field which distinguish between image space algorithms, hybrid algorithms and object space algorithms. In section 7.3, a new alternative visualization technique will be proposed, permitting to give interesting solutions to the scene understanding problem, for scenes containing lights, mirrors or transparent objects. In section 7.4, an alternative visualization method, permitting to understand a scene by seen simultaneously the outside of the scene and selected objects inside it, will be presented.

7.2 Main techniques for scene understanding

The very first works in the area of understanding virtual worlds were published at the end of 80's and the beginning of 90's. There were very few works because the computer graphics community was not convinced that this area was important for computer graphics. The purpose of these works was to offer the user a help to understand simple virtual worlds by computing a good point of view.

7.2.1 Best view computing for virtual worlds

When the virtual world to understand is simple enough, a single view of it may be enough to understand the virtual world. So, it is important to be able to propose an automatic computation of a "good" viewpoint.

Kamada and Kawai [19] consider a position as a good point of view if it minimizes the number of degenerated images of objects when the scene is projected orthogonally. A degenerated image is an image where more than one edges belong to the same straight line. They have proposed to minimize the angle between a direction of view and the normal of the considered plan for a single face or to minimize the maximum angle deviation for all the faces of a complex scene.

This technique is very interesting for a wire-frame display. However, it is not very useful for a more realistic display. Indeed, this technique does not take into account visibility of the elements of the considered scene and a big element of the scene may hide all the others in the final display.

Plemenos and Benayada [24] propose an iterative method of automatic viewpoint calculation and create a heuristic that extends the definition given by Kamada and Kawai. A point is considered as a good point of view if it gives; in addition of the minimization of deviation between a direction of view and normals to the faces, the most important amount of details. The viewpoint quality is based on two main geometric criteria: number of visible polygons and area of the projected visible part of the scene, it is computed using the following formula:

$$I(V) = \frac{\sum_{i=1}^{n} \left[\frac{P_i(V)}{P_i(V)+1}\right]}{n} + \frac{\sum_{i=1}^{n} P_i(V)}{r} \qquad (1)$$

Where: $I(V)$ is the importance of the view point V,

Pi(V) is the projected visible area of the polygon number i obtained from the point of view V,

r is the total projected area,

n is the total number of polygons of the scene.

[a] denotes the smallest integer, greater than or equal to a.

The process used to determine a good point of view works as follows:

The scene is placed on the center of the unit sphere whose surface represents all possible points of view. The sphere is divided into 8 spherical triangles (see Fig.7.1) and the best one will be whose vertices represent the best quality according to the formula (1). Finally the best point of view is computed by recursive subdivision on the best spherical triangle. (See Fig. 7.2)

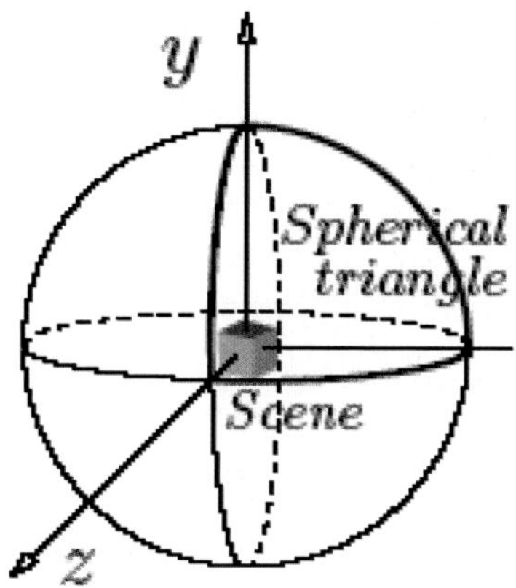

Fig. 7.1: Sphere subdivision in 8 spherical triangles

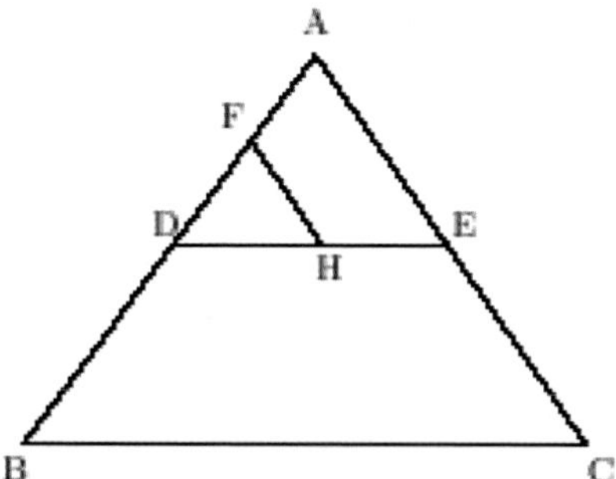

Fig. 7.2: Best spherical triangle recursive subdivision

This method gives generally interesting results. However, the number of polygons criterion may produces some drawbacks. To resolve this problem, Sokolov et al. [34, 35] propose to replace the number of polygon criterion by the criterion of *total curvature* of the scene which is given by the equation:

$$I(p) = \sum_{v \in V(p)} \left| 2\pi - \sum_{\alpha_i \in \alpha(v)} \alpha_i \right| \sum_{f \in F(p)} P(f) \tag{2}$$

Where:

      F(p) is the set of polygons visible from the viewpoint p,

      P(f) is the projected area of polygon f,

      V(p) is the set of visible vertices of the scene from p,

      $\alpha(v)$ is the set of angles adjacent to the vertex v.

Equation 2 uses the curvature in a vertex (see Fig. 7.3), which is the sum of angles adjacent to the vertex minus 2 .

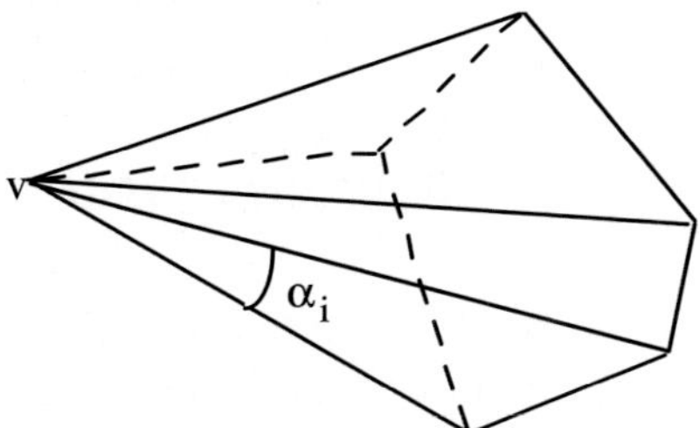

Fig. 7.3: Curvature in a vertex

The best viewpoint is computed by using a data structure, so-called *visibility graph*, which allows, to every discrete potential viewpoint on the surface of the surrounding sphere, the association of the visual pertinence of view from this viewpoint.

Colin [3] proposed a method to compute a good view for octree models. The viewpoint is considered to be good if it shows high amount of voxels. This method computes first the "best" initial spherical triangle and then the "best" viewpoint is approximately estimated on the chosen triangle.

Sbert et al. [32] use *viewpoint entropy* criterion to evaluate the amount of information captured from a given point of view which is defined by the following formula:

$$H_p(X) = - \sum_{i=0}^{N_f} \frac{A_i}{A_t} \log \frac{A_i}{A_t} \tag{3}$$

Where N_f is the number of faces of the scene, A_i is the projected area of the face i and A_t is the total area covered over the sphere.

The maximum entropy is obtained when a viewpoint can see all the faces with the same relative projected area Ai/At. The best viewpoint is defined as the one that has the maximum entropy.

When we have to understand a complex virtual world, the knowledge of a single point of view is not enough to understand it. Computing more than one point of view is generally not a satisfactory solution in most cases because the transition from a point of view to another one can disconcert the user, especially when the new point of view is far from the current one. The best solution, in the case of complex virtual worlds is to offer an automatic exploration of the virtual world by a camera that chooses good points of view and, at the same time, a path that avoids brusque changes of direction. Several authors have proposed methods for online or offline exploration [26, 27, 7, 38, 37, 16, 17, 34, 35].

7.2.2 Apparent contour extraction techniques

In computer graphics, contour extraction and rendering has a central role in a growing number of applications. It is not only used in non photo realistic rendering for artistic styles and cartoons, it is also used for technical illustrations, architectural design and medical atlases [13], in medical robotic [23], and for photo realistic rendering enhancement. It has been used as an efficient means to calculate shadow volumes [13], to rapidly create and render soft shadows on a plane [12]. It's also used to facilitate the haptic rendering [15]. Some authors, [5, 18], have described the use of silhouettes in CAD/CAM applications. Systems have also been built which use silhouettes to aid in modeling and motion capture tasks [1, 9, 20]. More applications and techniques based on line drawings detection are described in the course notes 7 of SIGGRAPH 05 [33].

Isenberg distinguish in his paper [14] image space algorithms, that only operate on image buffers, hybrid algorithms, that perform manipulations in object space but yield the silhouette in an image buffer, and object space algorithms, that perform all calculations in object space with the resulting silhouette represented by an analytic description of silhouette edges.

7.2.2.1 Image space algorithms

In image space, the easiest way to find significant lines would be by detecting discontinuities in the image buffer(s) that result from conventional rendering and extract them using image processing methods. This however, doesn't detect silhouette since changes in shading and texturing can cause erroneous edge detection. Saito and Takahachi [36], Decaudin [6], Hertzmann [11], Deussen and Strothotte [8], Nienhaus and Dellner [22], extend this method by using the geometric buffers known as G-Buffers which are the depth buffer (z-buffer), the normal buffer and the Id-buffer.

By detecting the 0 discontinuities of the depth map we can obtain the silhouette. And with the detection of the 1 discontinuities of the normal map we can obtain the crease edge. The advantage of image space algorithms is that they are relatively simple to implement because they operate with buffers which can be generated easily with the existing graphics hardware, can be applied to all kinds of models and give good results in simple cases. However, the main disadvantage of image space algorithms is that they depend on a threshold that has to be adjusted for each scene. A second disadvantage of these techniques is that in the case of complex images they might add undesired contours or miss some lines because of the presence of reflections and refractions.

7.2.2.2 Hybrid algorithms

Rustagi [30], Rossignac and Emmerik [29] use hybrid algorithms which are characterized by operations in object space (translations) that are followed by rendering the modified polygons in a more or less ordinary way using a z-buffer. This usually requires two or more rendering passes.

Raskar and Cohen [28], Gooch et al. [10], Raskar [31], they generalize this approach in their work by using traditional z-buffering along with back-face or front-face culling, respectively.

With hybrid algorithms, the visual appearance of the generated images tends to be a more stylistic one.

Similarly to image space algorithms, hybrid algorithm might fail to face some numerical problems due to limited z-buffer resolution.

7.2.2.3 Object space algorithms

The computation of silhouettes in this group of algorithms, as the name suggests, takes place entirely in object space. The trivial object space algorithm is based on the definition of a silhouette edge. The algorithm consists of two basic steps. First, it classifies all the mesh polygons as front or back facing, as seen from the camera. Next, the algorithm examines all model edges and selects only those that share exactly one front- and one back-facing polygon. The algorithm must complete these two steps for every frame.

Buchanan and Sousa [2], suggest, to support this process, the use of a data structure called an *edge buffer* where they store two additional bits per edge, F and B for front and back facing. They extend later this idea to support border edges and other feature edges.

This simple algorithm, with or without using the edge buffer data structure, is guaranteed to find all silhouette edges in the model. It is easy to implement and well suited for applications that only use small models. However, it is computationally expensive for common hardware-based rendering systems and the model sizes typically used with them. It must look at every face, determine face orientation (using one dot product per face; for perspective projection it must re-compute the view vector for every face), and look at every edge. This is a linear method in terms of the number of edges and faces but too slow for interactive rendering of reasonably sized models.

To speed up the algorithm, Card and Mitchell [4] suggest employing user-programmable vertex processing hardware (shaders).

Hertzmann and Zorin [13] consider the silhouette of a free-form surface approximated by a polygonal mesh. To find this silhouette, they re-compute the normal vectors of the approximated free-form surface in the vertices of the polygonal mesh. Using this normal, they compute its dot product with the respective viewing direction. Then, for every edge where the sign of the dot product is different at both vertices, they use linear interpolation along this edge to find the point where it is zero. These points connect to yield a piecewise linear sub-polygon silhouette line. The resulting silhouette line is likely to have far fewer artifacts. Also, the result is much closer to the real silhouette than the result of a traditional polygonal silhouette extraction method.

Nehab and Gattas [21] propose a completely different approach for ray traced images. They create an algorithm that divides ray into equivalence categories or classes. They consider a pixel as representing the ray cast through its lower left corner. To determine the pixels that spawn edges, each pixel is compared against its three 4-neighbors (a, b, c for pixel p in Fig. 7.4). If any differences are detected, the pixel is selected as an edge pixel.

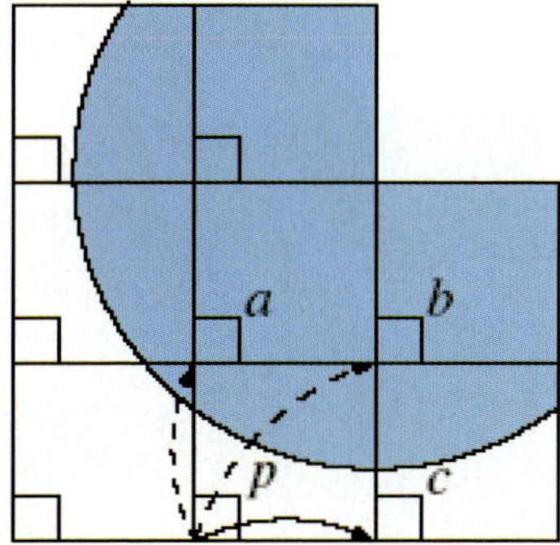

Fig. 7.4: Edges are detected when a pixel's category is different from that of one of its neighbors.

The path followed by a ray is represented by a tree (see Fig. 7.5) and can be described by:

1.   The ray who intersects the object
2.   Rays coming from lights visible by the intersected object
3.   Reflected & refracted rays

   Two pixels are in the same category if their ray trees are equal (have the same word obtained by joining the summit of Knots):

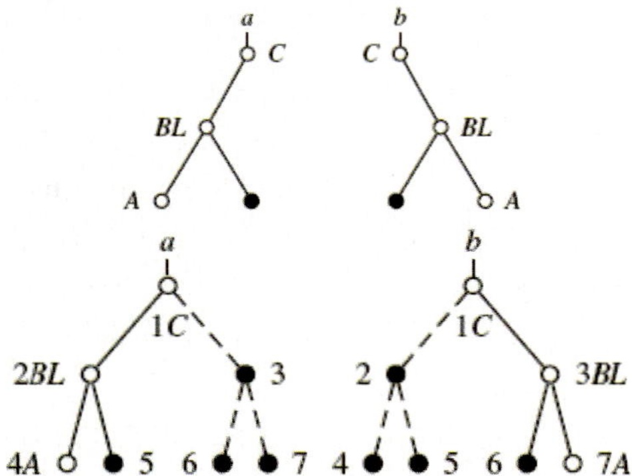

Fig. 7.5: With the numbering of the node positions, trees a and b receive strings 4A2BL1C and 7A3BL1, respectively.

The results are very good for simple scenes. However, for complex scenes, the method can fail unpredictably specially with scenes that represent reflections and refractions; refracted and reflected object's edges are detected (see Fig. 7.6)

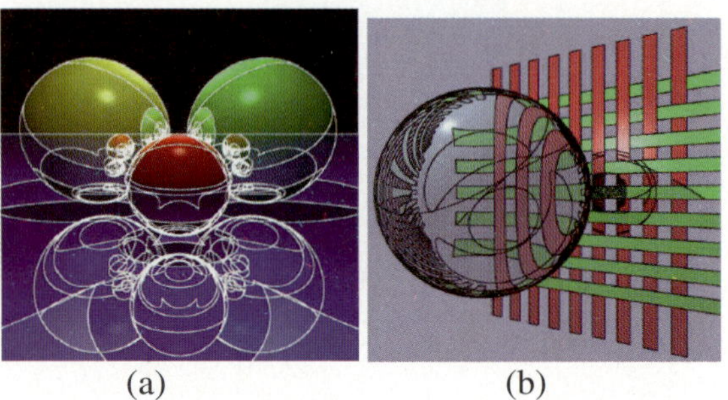

(a)   (b)

Fig. 7.6: (a) A highly reflective scene. (b) Refraction example.

7.3 Understanding visually complex scenes

In this section, we present a new approach to face the problem of understanding scenes containing lights mirrors and transparent objects. The proposed method is based on ray casting and selective refinement improvement, in order to extract apparent contours of the real objects of the scene.

The term of Non Photorealistic Rendering (NPR) was, a long time ago, only applied for artistic illustrations such as Pen and Ink or watercolor. Many computer graphics researchers are nowadays exploring NPR techniques as an alternative to realistic rendering. More importantly, NPR is now being acknowledged for its ability to communicate the shape and structure of complex scenes. These techniques can emphasize specific features, convey material properties and omit extraneous or erroneous information. Therefore, the method presented in this section is going to imitate and take inspiration from these techniques by extracting apparent contours of real objects present in the scene in order to make a better knowledge through realistic rendering.

We define a visually complex scene as a scene containing lights, mirrors and transparent objects. Such a scene is sometimes difficult to understand with a realistic rendering. To face this problem we propose to delimit the objects of the scene by their apparent contour before rendering the image and before adding transparency, reflections and/or refractions. With this technique, the real objects of the scene may be overlaid on the realistic rendering, making the user able to distinguish reality from reflection and refraction effects.

Our approach is divided in two parts:

1. The selective refinement part [25], that searches for an initial contour pixel related to each real object presents in the scene.
2. The code direction part that searches for the complete contours.

Both parts use simplified AI heuristic search.

7.3.1 The selective refinement part

Our goal is to find for each object, one contour point which will be used as a departure point for our code direction method.

First, we divide the image of pixels into a set of macro pixels (8x8 pixels). For each macro pixel, we send rays to the up right (UR), up left (UL) down right (DR) and down left (DL) pixels to detect for each ray the ID of the closest object. We associate each returned ID to its correspondent pixel.

The macro pixels which represent different intersections must contain a contour pixel. They are considered as our useful macro pixels which are subdivided into 4 sub macro pixels. The same process is applied to each sub macro pixel until we obtain a block of 2x2 pixels (see Fig. 7.7). The block of 2x2 pixels that has intersection with different objects, contains certainly at least a one contour point. More we have different intersections in the block, more we have initial contour pixels. To avoid having more than one initial contour pixel for the same object, since we get the first contour pixel of an object, we neglect all other pixels that have the same ID.

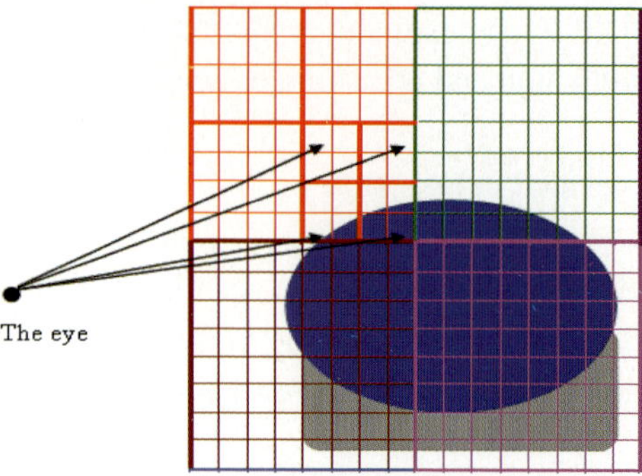

Fig. 7.7: The macroPixel that intersects different objects is divided into 4 sub
macro pixels.

7.3.2 The code direction part

The code direction algorithm starts with an initial contour pixel and follows, at
each time, a certain direction that conducts to the following contour point. The
same process is repeated until the complete contour of an object was obtained.

In other words, this method can be applied by following these 3 steps:

1.   Choose the starting point.
2.   Choose the initial direction to the second contour point.
3.   Choose the following direction to the following contour point

Before speaking about the steps of the algorithm, we have to define first for
each pixel, its 8 neighbors. Each pixel in the neighborhood has a previous and a
following pixel respecting the order indexed from 0 to 7. For example the
neighbor number 1 (coordinate(x-1, y+1)) has the neighbors 0 (coordinate (x,
y+1)) and 2 (coordinate(x-1, y)) as its previous and following pixels (see Fig. 7.8)

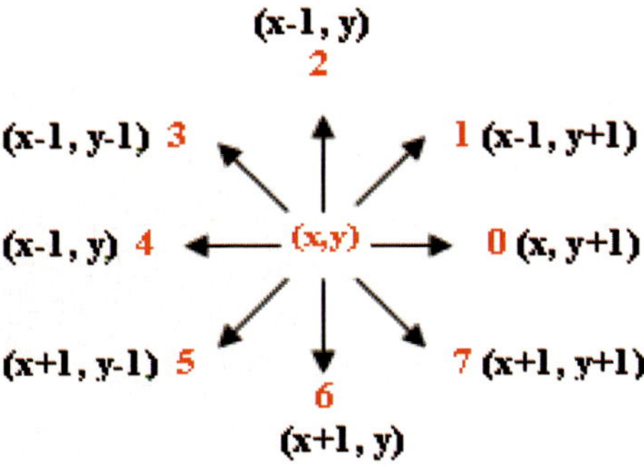

Fig 7.8: The coordinates of the 8 neighbors indexed from 0 to 7

Step 1: Choose the starting point

The algorithm starts with the initial contour pixels obtained by the selective refinement method.

More there are initial pixels, obtained by the selective refinement, more there are contours that have to be detected.

Step 2: Choose the initial direction to the second contour point

The second contour pixel should be one of the 8 neighbors of the starting one.

A ray is sent to each neighbor. The first one that has the same ID as the starting point will be our current pixel that has to be tested as contour or non contour point.

1. The current pixel is considered as a contour point if its previous and following pixels don't have the same ID.

2. If the tested pixel wasn't a contour point, we pass to the other neighbors until we get the one that has the same ID of the departure point and apply the same test.

3. If all the neighbors were tested and none of them was a contour pixel, contour extraction is stopped.

Step 3: Choose the following direction to the following contour point

Since a second contour pixel is found, it is colored with a special color (red) and the same process is applied to find the following direction by considering the obtained contour pixel as a starting one. In order to avoid a return to a chosen contour point, we only test between the 8 neighbors which are not colored yet.

The algorithm will stop when one of the following two cases is encountered:

1.   Return to the initial starting point (closed contour)
2.   None of the neighbors of the current pixel is a contour point (open contour)

7.3.3 First results

The method presented in this section has been implemented and obtained results allow to conclude that it is possible to visually enhance the understanding of complex scenes which contain reflections, refractions and shadows by detecting real object contours.

The apparent contours are traced in special color (currently red) in order to overlay the real objects present in the scene. Results are shown in Fig. 7.9 and timing results for the contours computation cost and the rendering cost are given in the Table 1. The images shown in Fig. 7.9 represent different scene cases.

Scenes 1 and 2 represent the same objects with and without reflections and refractions respectively. For both of them the same contour has been obtained and with the same cost in time. This is due to the fact that the algorithm computes the contour directly from the scene, not from its image, and it doesn't depend on the way it will be rendered. Moreover, Scene 1 is not visually well understood because of the presence of the shadow in the floor and its reflection on the sphere. The detection of contours makes the visualization better and helps the observer to distinguish the 3 spheres from, shadows, reflections and refractions.

Scene 3 represents objects which hide others, whereas Scene 4 represents the case of objects in shade where objects intersect each other. In both of them, the presented method is able to detect visible silhouettes.

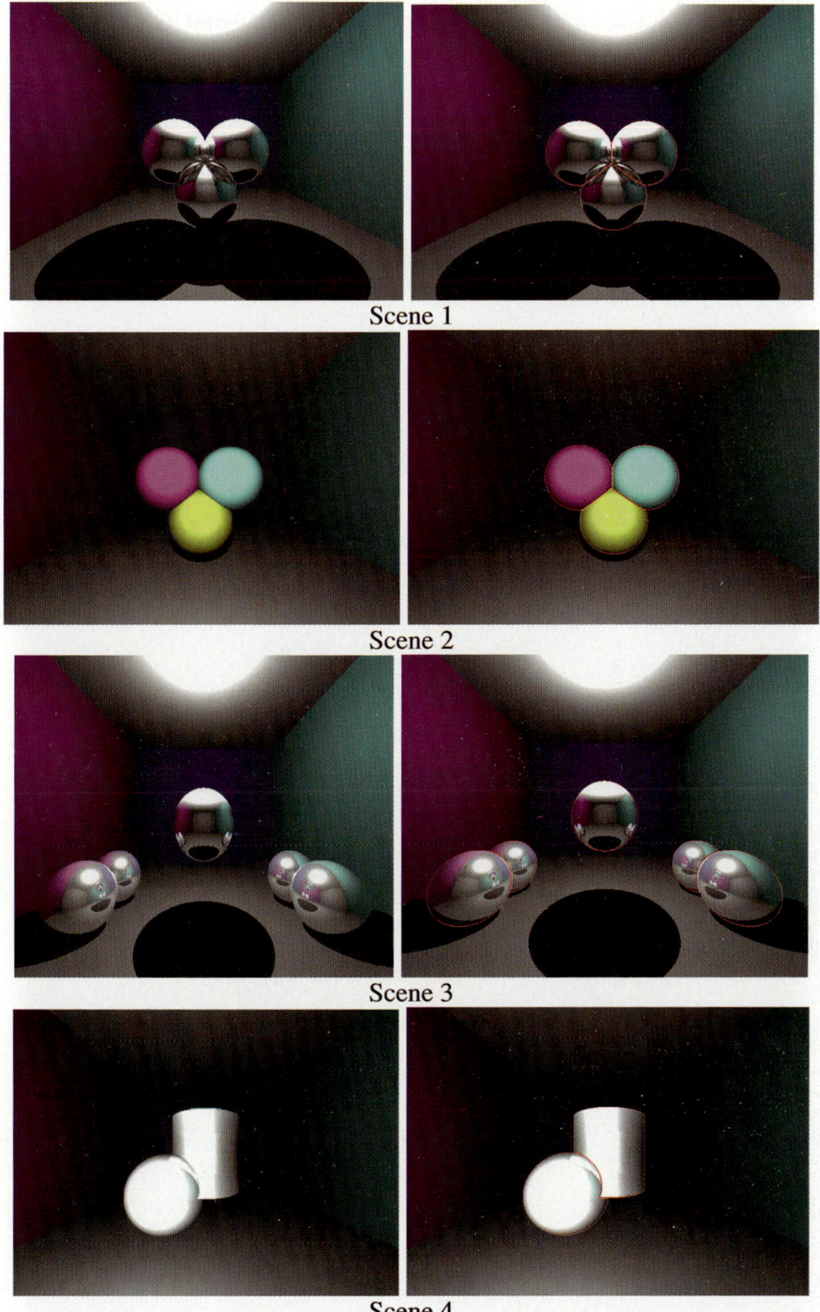

Fig. 7.9: Visual results for four test scenes. On the left, realistic rendering of the scene; on the right, apparent contour drawing (Images Nancy Dandachy)

Fig. 7.10 shows realistic rendering of another scene, followed by the same rendering, but with contour drawing.

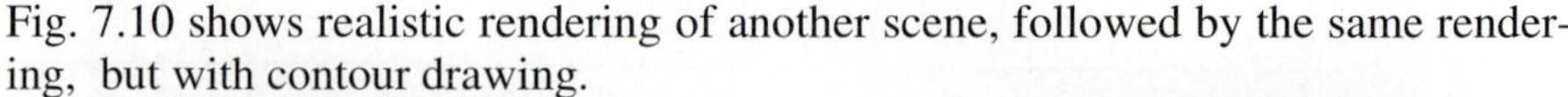

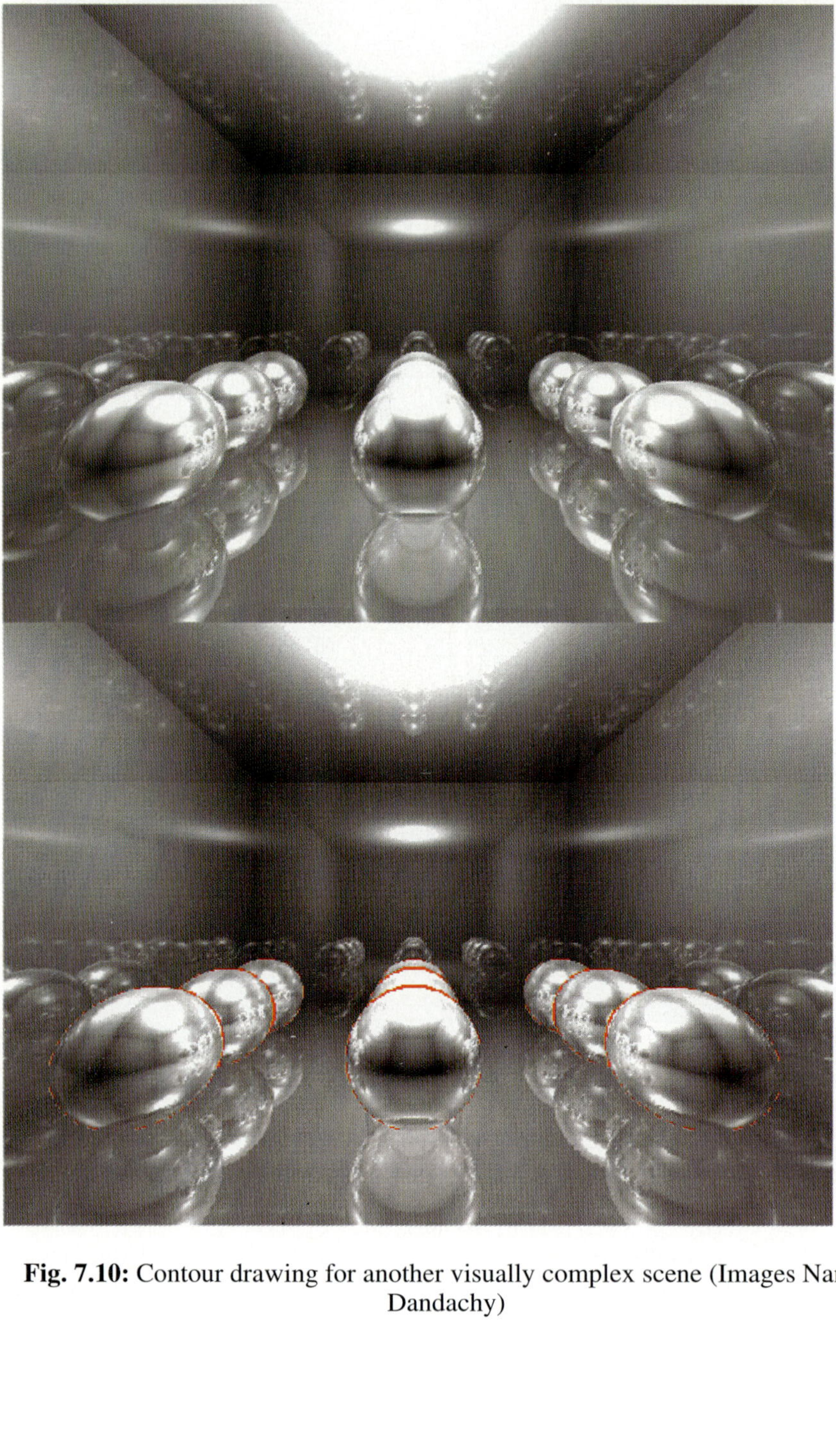

Fig. 7.10: Contour drawing for another visually complex scene (Images Nancy Dandachy)

Scene	Time to render the scene (seconds)	Time to detect the contour (seconds)
Scene 1	16.122	0.15
Scene 2	15.172	0.15
Scene 3	17.806	0.311
Scene 4	17.806	0.1

Table 1: Time results for rendering each scene and detecting its contour

The method works well for any scene made of curved surfaces. However, in its current version, this method fails in the case of polygonal meshes, because it detects the contour of each polygon of the mesh. In Fig. 7.11 one can see contour extraction of a sphere modeled by a set of polygons.

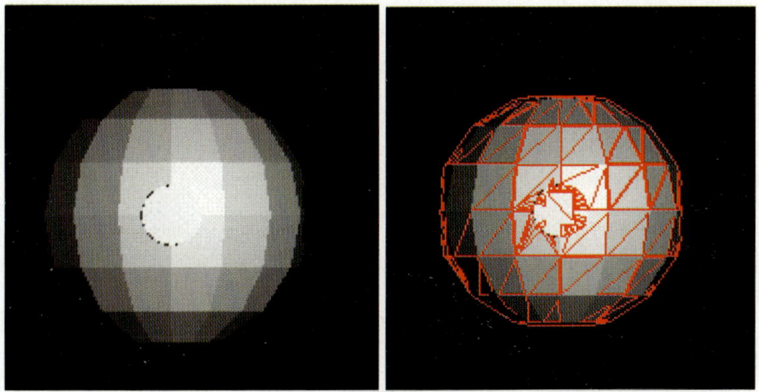

Fig. 7.11: The method fails in the case of polygonal meshes (Images Nancy Dandachy)

7.3.4 Discussion

The scene understanding method presented in this section is a new method permitting to understand visually complex scenes. This kind of methods can be used to improve scene understanding in the case of visually complex realistically rendered scenes, comporting mirrors and light effects.

The proposed method combines ray casting and selective refinement and allows extracting the apparent contours of the "real" objects of a scene. These contours, overlaid on the realistic rendering of the scene, allow the user to distinguish between parts of the image that correspond to real objects of the scene and the reflections, refractions and shadows parts. The first obtained results seem convincing for scenes made of curved surfaces. However, the method has to be improved, in order to include objects made of several patches.

This method, could be combined with techniques computing a good point of view and highly improve visually complex scene understanding.

7.4 Understanding scenes with hidden details

Sometimes, in newspapers, one can read descriptions of new aircraft models, or other kinds of vehicles, spatial stations and so on. Very often, journalists use special illustrations of their descriptions with pictures where it is possible to see the vehicle from outside and, at the same time, to see some details inside the vehicle, thanks to holes designed on the picture. This is a way to give the lectors a general idea of the whole vehicle. In figure 7.12, one can see a (bad) example of this kind of illustrations. It is the only we have had the reflex to keep.

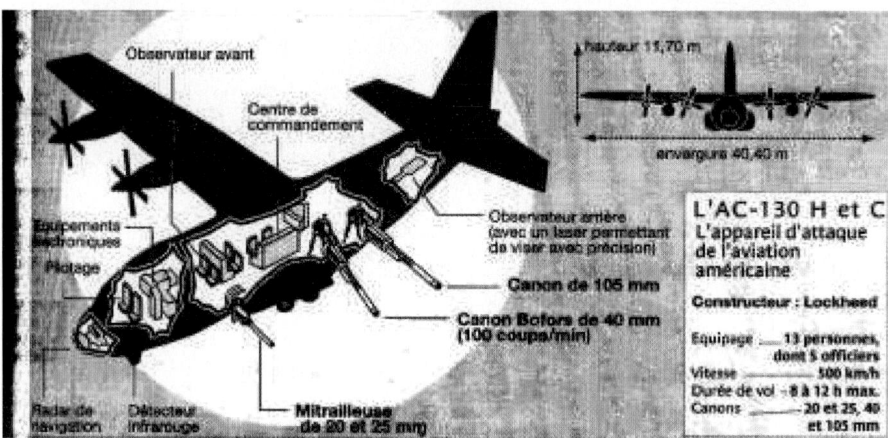

Fig. 7.12: A manually illustrated aircraft, where details inside the aircraft are visible from outside

Of course, all these illustrations are manually designed and it is a very difficult work. Moreover, this work is to be done for each new illustration. It is obvious that such a work could be made automatically by a program, for any vehicle or other object, if an internal model of this object exists.

The general question to answer is: how to render a scene in order to see simultaneously its outside and non visible objects inside the scene.

7.4.1 Using different rendering modes

A first naive idea could be to use two different visualization modes for the outside of the scene and for the objects inside it [39]. In Fig. 7.13, the teapot inside the sphere is rendered with hidden surfaces removed, whereas for the sphere wire frame rendering is used. On the of Fig. 7.13, additional back facing culling (BFC) was performed in order to improve the image quality.

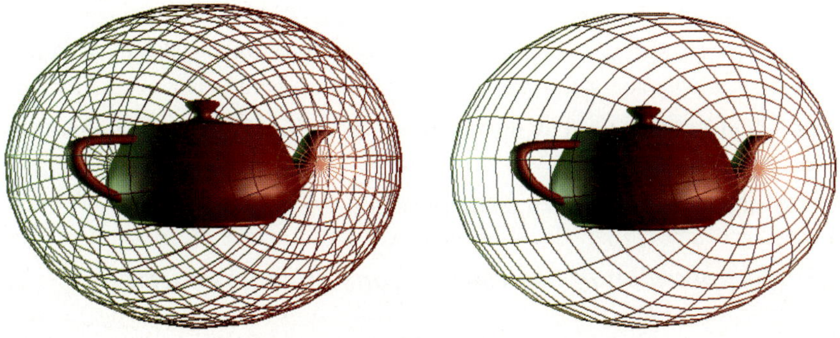

Fig. 7.13: Two different rendering modes for the objects of a scene (Images Nancy Dandachy)

This solution has too many drawbacks. First, if the external surface of the scene contains many details, it is impossible to understand them with a wire frame rendering. Moreover, all objects inside the scene are visible, not only selected ones.

7.4.2 Creating holes on the external surfaces of the scene

A better idea is to create a hole proportional to the size of the objects inside the scene we want to be visible [39]. More precisely, the purpose is to suppress all polygons which may hide the selected objects inside the scene.

There are two methods to suppress useless polygons of the scene. The first one needs a preprocessing step, where useless polygons are suppressed. After that, the

scene is normally rendered. The second method processes useless polygons during
the rendering of the scene.

7.4.2.1 Using a preprocessing step

The first step of the method is used to suppress useless polygons, that is polygons
hiding interesting inside objects which the user would like to see. This step works
as follows:

1. The bounding box of each object, or the bounding box of the set of objects, to
   be visible is defined.
2. The defined bounding box(es) is (are) projected to the image space.
3. A ray is sent towards each pixel covered by the projection of the bounding
   box(es).
4. All polygons intersected by the ray before the selected objects are removed.

The next step is rendering. Any rendering algorithm may be used.

7.4.2.2 Suppression of useless polygons during rendering

Here, there is not any preprocessing. A modified rendering algorithm has to be
used, which has to take into account the presence of possible useless polygons,
that is polygons which hide parts of objects to be seen.

This method is less interesting that the first one because it cannot work with traditional rendering algorithms and needs to modify them.

7.4.2.3 Some results

The first results obtained with the technique presented in this subsection seem interesting enough. In Fig. 7.14, one can see an example of scene understanding by
creating holes on the scene. In our case, the user can the small sphere covered by
the big one. To make it possible, polygons have been automatically removed from
the front part of the big sphere.

However, in some cases, results are not very satisfactory, especially in the case
where the scene contains big size polygons. In such a case, obtained holes are too
large and the visual impression is bad. This problem is illustrated n Fig. 7.15,

where too big polygons have been removed, due to the big size of some polygons of the scene

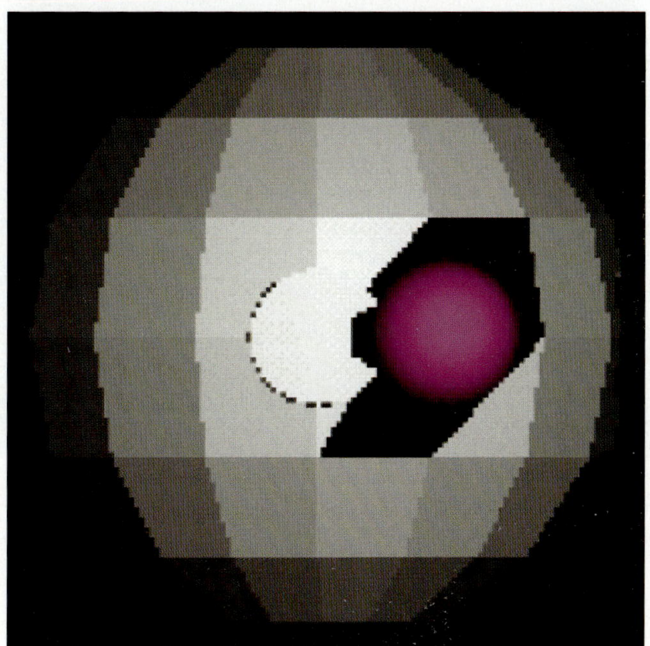

Fig. 7.14: Several polygons have been automatically removed from the big sphere, allowing the user to see the small spere located inside the big one (Image Nancy Dandachy)

Fig. 7.16 and Fig. 7.17 show results obtained with another technique, where the all the objects of the scene, except for the useful ones, are rendered first. Then, the useful objects, that is the objects inside the scene that the user would like to see, are rendered and superposed on the image. Finally, the pixels around the silhouette of the useful objects are darkened.

Fig. 7.15: Ugly results may be obtained when the scene contains big polygons (Image Nancy Dandachy)

Fig. 7.16: A small sphere inside a big one (Image Nancy Dandachy)

Fig. 7.17: A sphere inside a kind of house (Image Nancy Dandachy)

This last solution is not satisfactory for three reasons:

1. It is difficult to detect the silhouettes of a set of objects with possible intersections.
2. Darkening sihouettes may darken parts of useful objects too.
3. Seeing objects inside other objects is interesting only when it permits so see not only these objects but also their context in the scene. Otherwise, the result is visually confusing and does not permit to well understand the scene.

We think that the best and simplest solution is to improve the method of automatic suppression of front polygons, in order to avoid the drawbacks we have seen above.

A simple method to improve results could work, in two steps, as follows:

First step: Preprocessing

1. The bounding box of each object, or the bounding box of the set of objects, to be visible is defined.
2. The defined bounding box(es) is (are) projected to the image space.
3. A ray is sent towards each pixel covered by the projection of the bounding box(es).

4. All polygons intersected by the ray before the selected objects and whose
   height or width is greater than a threshold value, are subdivided in a set of
   smaller polygons (or triangles) and replace the initial polygons.
5. A ray is sent again towards each pixel covered by the projection of the bounding box(es).
6. All polygons intersected by the ray before the selected objects are removed.

Second step: Rendering:

Any rendering algorithm may be used.

This improvement is not yet implemented. Many other improvements are possible
as, for example, combination with other scene understanding techniques, permitting to compute a good point of view and taking into account the possibility to
create holes on the scene.

7.5 Discussion

The goal of the techniques presented in this chapter was to show that the problem
of scene understanding is a difficult one and, if efficient methods, using Artificial
Intelligence techniques, exist, there are special cases where these methods may
fail. In such cases, the human intelligence may replace Artificial Intelligence and
propose solutions, applicable to the current situation, in order to help the user understand scenes in this special context. This is a work on the data visualization
area, where data are scenes.

Two kinds of techniques were presented, corresponding to two particular cases of
scenes, where understanding by general scene understanding methods may fail.
The first case is the case where combination of light effects produces visually
complicated images, which don't help the user understand the corresponding
scenes. Another particular case, the second one, is when a scene has to be understood from only one image and, for a good understanding, we need to see at the
same time the external part of it and selected parts inside the scene.

We think that the solutions presented here are interesting, even if they are not entirely satisfactory. They are only the first steps of this interesting area. We are sure
that, based on the presented techniques, more elaborated solutions may be given to
these particular cases. In section 7.4 we have suggested some improvements and

we think that there are many open problems in this research area. Moreover, it is possible to discover many other particular cases, where special solutions have to be imagined for efficient scene understanding. Of course, these solutions may be combined with general scene understanding techniques.

References

1. Bottino, A., Laurentini, A.: Experimenting with non instructive motion capture in a virtual environment. The visual Computer 17(1), 14–29 (2001)
2. Buchanan, J.W., Sousa, M.C.: The Edge Buffer: A Data Structure for Easy Silhouette Rendering. In: Proceedings 1st Int'l Symp. Non-Photorealistic Animation and Rendering, pp. 39–42. ACM Press, New York (2000)
3. Colin, C.: A System for Exploring the Universe of Polyhedral Shapes. In: Eurographics 1988, Nice, France (September 1988)
4. Card, D., Mitchell, J.L.: Non-Photorealistic Rendering with Pixel and Vertex Shaders. In: Engel, W. (ed.) Vertex and Pixel Shaders Tips and Tricks, Wordware (2002)
5. Chung, Y.C., Park, J.W., Shin, H., Choi, B.K.: Modeling the surface swept by generalized cutter for NC verification. Computer-aided Design 30(8), 587–594 (1998)
6. Decaudin, P.: Cartoon looking rendering of 3D scenes. Research Report INIRIA 2919 (June 1996)
7. Dorme, G.: Study and implementation of 3D scene understanding techniques. PhD thesis, University of Limoges, France (June 2001) (in French)
8. Deussen, O., Strothotte, T.: Computer-Generated Pen and Ink Illustration of Trees. In: Proceedingss Siggraaph 2000, Computer Graphics. Proceedings Ann. Conf. Series, vol. 34, pp. 13–18. ACM Press, New York (2000)
9. Fua, P., Plankers, R., Thalmann, D.: From synthesis to analysis: Fitting human animation models to image data. In: Computer Graphics Internationnal 1999, p. 4. IEEE CS Press, Los Alamitos (1999)
10. Gooch, B., et al.: Interactive Technical Illustration. In: Proceedings 1999 ACM Symp. Interactive 3D Graphics, pp. 31–38. ACM Press, New York (1999)
11. Hertzmann, A.: Introduction to 3D Non-Photorealistic Rendering: Silhouettes and outlines. In: Green, S. (ed.) Non-Photorealistic rendering (Siggraph 1999 Course Notes). ACM Press, New York (1999)
12. Haines, E.: Soft planar shadows using plateaus. Journal of graphics Tools 6(1), 19–27 (2001)
13. Hertzmann, A., Zorin, D.: Illustrating Smooth Surfaces. In: Spencer, S.N. (ed.) Proceedings Siggraph 2000, Computer Graphics. Proceedings Ann. Conf. Series, pp. 517–526. ACM Press, New York (2000)
14. Isenberg, T., et al.: A Developper's Guide to silhouette Algorithms for Polygonal Models. IEEE Computer Graphics and Applications 23(4), 28–37 (2003)
15. Johnson, D.E., Cohen, E.: Spatialized normal cone hierarchies. In: Symposium on interactive 3D Graphics, March 1, pp. 129–134. ACM, New York (2001)
16. Plemenos, D., Grasset, J., Jaubert, B., Tamine, K.: Intelligent visibility-based 3D scene processing techniques for computer games. In: GraphiCon 2005, Novosibirsk, Russia (June 2005)
17. Jaubert, B., Tamine, K., Plemenos, D.: Techniques for off-line exploration using a virtual camera. In: International Conference 3IA 2006, Limoges, France, May 23–24 (2006)
18. Jensen, C.G., Red, W.E., Pi, J.: Tool selection for five axis curvature matched machining. Computer-aided Design 34(3), 251–266 (2002)

19. Kamada, T., Kawai, S.: A Simple Method for Computing General Position in Displaying Three-dimensional Objects. Computer Vision, Graphics and Image Processing 41 (1988)
20. Lee, W., Gu, J., Magnenat-Thalmann, N.: Generating animable 3D virtual humans from photographs. Computer Graphics Forum 19(3) (2000) ISSN 1067-7055
21. Nehab, D., Gattas, M.: Ray Path Categorization. In: Proceedingss of the Brazilian Symposium on Computer Graphics and Image Processing -SIBGRAPI, Gramado, Brazil, pp. 227–234 (2000)
22. Nienhaus, M., Doellner, J.: Edge Enhancement- An algorithm for real time Non-Photorealistic Rendering. Journal of WSCG 2003 11(1) (2003) ISSN 1213-6972
23. Olson, M., Zhang, H.: Silhouette Extraction in Hough Space. Computer Graphics Forum 25(3), 273–282 (2006) (special issue on Eurographics 2006)
24. Plemenos, D., Benayada, M.: Intelligent display in scene modeling. New techniques to automatically compute good views. In: GraphiCon 1996, Saint Petersburg (July 1996)
25. Plemenos, D.: A contribution to the study and development of scene modelling, generation and visualisation techniques. The MultiFormes project, Professorial dissertation, Nantes, France (November 1991)
26. Barral, P., Dorme, G., Plemenos, D.: Visual understanding of a scene by automatic movement of a camera. In: GraphiCon 1999, Moscow, Russia, August 26 - September 3 (1999)
27. Barral, P., Dorme, G., Plemenos, D.: Scene understanding techniques using a virtual camera. In: Short paper, Eurographics 2000, Interlagen, Switzerland, August 20 - 25 (2000)
28. Raskar, R., Cohen, M.: Image Precision Silhouette Edges. In: Spencer, S.N. (ed.) Proceedings 1999 ACM Symp. Interactive 3D Graphics, pp. 135–140.11. ACM Press, New York (1999)
29. Rossignac, J.R., Van Emmerik, M.: Hidden Contours on a Frame-Buffer. In: Proceedings 7th Eurographics Workshop Computer Graphics Hardware, Eurographics, pp. 188–204 (1992)
30. Rustagi, P.: Silhouette Line Display from Shaded Models, Iris Universe, pp. 42–44 (Fall 1989)
31. Raskar, R.: Hardware Support for Non-Photorealistic Rendering. In: Proceedings 2001 Siggraph/Eurographics Workshop onGraphics Hardware, pp. 41–46. ACM Press, New York (2001)
32. Sbert, M., Feixas, M., Rigau, J., Castro, F., Vazquez, P.-P.: Applications of the information theory to computer graphics. In: International Conference 3IA 2002, Limoges, France, May 14-15 (2002)
33. Prusinkiewicz, S., et al.: Line Drawings from 3D Models. In: International Conference on Computer Graphics and Interactive Techniques, ACM Siggraph 2005 Course 7, Los Angelos, California, vol. 1 (July 2005)
34. Sokolov, D., Plemenos, D.: Viewpoint quality and scene understanding. In: VAST 2005 Eurographics Symposium Proceedingss, Pisa, Italy, pp. 67–73 (2005)
35. Sokolov, D., Plemenos, D., Tamine, K.: Methods and data structures for virtual world exploration. The Visual Computer (2006)
36. Saito, T., Takahashi, T.: Comprehensible Rendering of 3-D Shapes. Computer Graphics (SIGGRAPH 1990 Proceedings) 24, 197–206 (1990)
37. Vazquez, P.-P.: On the selection of good views and its application to computer graphics. PhD Thesis, Barcelona, Spain, May 26 (2003)
38. Vazquez, P.-P., Sbert, M.: Automatic indoor scene exploration. In: Proceedingss of the International Conference 3IA 2003, Limoges, France, May 14-15 (2003)
39. Dandachy, N.: Alternatine visualization techniques for 3D scene understanding. Ph. D. Thesis, Limoges, France, November 20 (2006)
40. Dandachy, N., Plemenos, D., El Hassan, B.: Scene understanding by apparent contour extraction. In: International Conference 3IA 2007, Athens, Greece, May 30-31 (2007)

Printing: Krips bv, Meppel, The Netherlands
Binding: Stürtz, Würzburg, Germany